Progress in Probability and Statistics
Vol. 1

Edited by
Peter Huber
Murray Rosenblatt

Birkhäuser
Boston · Basel · Stuttgart

Seminar on Stochastic Processes, 1981

E. Çinlar,
K.L. Chung,
R.K. Getoor,
editors

1981

Birkhäuser
Boston • Basel • Stuttgart

Editors:

E. Çinlar
Department of Mathematics
Northwestern University
Evanston, Illinois 60201

K.L. Chung
Department of Mathematics
Stanford University
Standford, California 94305

R.K. Getoor
Department of Mathematics
University of California
La Jolla, California 92093

LC 81-71608

ISBN 3-7643-3072-4
Printed in USA

TABLE OF CONTENTS

FOREWORD

This volume consists of about half of the papers presented during a three-day seminar on stochastic processes held at Northwestern University in April 1981. The aim of the seminar was to bring together a small group of kindred spirits working on stochastic processes and to provide an informal atmosphere for them to discuss their current work. We plan to hold such a seminar once a year, with slight variations in emphasis to reflect the changing concerns and interests within the field.

The invited participants in this year's seminar were J. AZEMA, R.M. BLUMENTHAL, R. CARMONA, K.L. CHUNG, R.K. GETOOR, J. JACOD, F. KNIGHT, S.OREY, A.O. PITTENGER, J. PITMAN, P. PROTTER, M.K. RAO, M. SHARPE, and J. WALSH. We thank them and other participants for the productive liveliness of the seminar. As mentioned above, the present volume is only a fragment of the work discussed at the seminar, the other papers having been already committed to other publications.

The seminar was made possible through the enlightened support of the Air Force Office of Scientific Research, Grant No. 80-0252. We are grateful to them as well as the publisher, Birkhäuser Boston, for their support and encouragement.

E.Ç.

Evanston, 1981

FEYNMAN-KAC FUNCTIONAL AND THE SCHRÖDINGER EQUATION*

by

K.L. CHUNG and K.M. RAO

The Feynman-Kac formula and its connections with classical analysis were inititated in [3]. Recently there has been a revival of interest in the associated probabilistic methods, particularly in applications to quantum physics as treated in [7]. Oddly enough the inherent potential theory has not been developed from this point of view. A search into the literature after this work was under way uncovered only one paper by Khas'minskii [4] which dealt with some relevant problems. But there the function q is assumed to be nonnegative and the methods used do not apparently apply to the general case; see the remarks after Corollary 2 to Theorem 2.2 below.** The case of q taking both signs is appealing as it involves oscillatory rather than absolute convergence problems. Intuitively, the Brownian motion must make intricate cancellations along its paths to yield up any determinable averages. In this respect Theorem 1.2 is a decisive result whose significance has yet to be explored. Next we solve the boundary value problem for the Schrödinger equation $(\Delta+2q)\varphi = 0$. In fact, for a

*Research supported in part by a NSF Grant MCS-8001540.

**The case $q \le 0$ is "trivial" in the context of this paper. For this case in a more general setting see [9, Chapter 13].

positive continuous boundary function f, a solution is obtained in the explicit formula given in (2) of §1 below, provided that this quantity is finite (at least at one point x in D). Thus the Feynman-Kac formula supplies the natural Green's operator for the problem. For a domain with finite measure, the result is the best possible as it includes the already classical solution of the Dirichlet problem by probability methods. Other results are valid for an arbitrary domain and it seems that some of them are proved here under less stringent conditions than usually given in non-probabilistic treatments. For instance, no condition on the smoothness of the boundary is assumed beyond that of regularity in the sense of the Dirichlet problem, and the basic results hold without this regularity. Of course, the Schrödinger equation is a case of elliptic partial differential equations on which there exists a huge literature, but we make no recourse to the latter theory. Comparisons between the methods should prove worthwhile and will be discussed in a separate publication.

It is well known that the Schrödinger equation differs essentially from the Laplace equation in that a condition on the size of the domain is necessary to guarantee the uniqueness of solution. In our context it is evident at the outset that the key to this is the quantity $u_D(x) = E^x\{ \exp(\int_0^{\tau_D} q(x(t))\, dt) \}$, the finiteness of which lies at the base of the probabilistic considerations. As natural as it is from our point of view, this quantity does not lend itself easily to non-probabilistic analysis. The identification in the simplest case (see the remark after Lemma D) as a particular solution of the equation is one of those amusing twists not uncommon in other theories when dealing with an object which has really a simple probabilistic existence.

The one-dimensional case of this investigation has appeared in [1] though the orientation is somewhat different there. A summary of the present results has been announced in [2].

1. Harnack inequality; global bound; boundary limit

Let $\{ X(t), t \geq 0 \}$ be the Brownian motion process in R^d, $d \geq 1$; with all paths continuous. The transition semigroup is $\{ P_t, t \geq 0 \}$ and $\mathcal{F}_t$ is the σ-field generated by $\{ X_s, 0 \leq s \leq t \}$ and augmented in the usual way. The qualifying phrase "almost surely" (a.s.) will be omitted when readily understood. A "set" is always a Borel set and a "function" is always a Borel measurable function. The class of bounded functions will be denoted by $b\mathcal{B}$; if its domain is A this is indicated by $b\mathcal{B}(A)$. Similarly for other classes of functions to be used later. The sup-norm of $f \in b\mathcal{B}$ is denoted by $\|f\|$; restricted to A it is denoted by $\|f\|_A$. P^x and E^x denote the probability and expectation for the process starting at x.

For any set B we put

$$\tau(B) = \tau_B = \inf\{ t > 0 \mid X(t) \notin B \} \quad ;$$

namely the first exit time from B, with the usual convention that $\inf \phi = \infty$. Let $q \in b\mathcal{B}$; as an abbreviation we put

(1) $$e_q(t) = \exp\{ \int_0^t q(X(s))\, ds \} \quad ;$$

when q is fixed it will be omitted from the notation. A domain in R^d is an open connected set; its boundary is $\partial D = \bar{D} \cap \overline{D^c}$, where $\bar{D}$ is the closure and D^c the complement of D. For $f \geq 0$ on ∂D we put for $x \in \bar{D}$:

(2) $$u(q,f;x) = E^x\{ e_q(\tau_D)\, f(X(\tau_D));\ \tau_D < \infty \}.$$

The following result is a case of Harnack's inequality, on which there is a considerable literature for elliptic partial differential equations.

Theorem 1.1. Let D be a domain and K a compact subset of D. There exists a constant $A > 0$ which depends only on D, K and Q, such that for any q with $\|q\| \le Q$ and $f \ge 0$ such that $u(q,f;\cdot)\not\equiv\infty$ in D, we have for any two points x_1 and x_2 in K:

$$A^{-1}\, u(q,f;x_2) \le u(q,f;x_1) \le A\, u(q,f;x_2) \quad . \tag{3}$$

Proof. We write $u(x)$ for $u(q,f;x)$. By hypothesis there exists $x_0 \in D$ such that $u(x_0) < \infty$. We may suppose $x_0 \in K$ by enlarging K. For any $r > 0$ define

$$T(r) = \inf\{\, t > 0 \mid \rho(X(t), X(0)) \ge r \,\}$$

where ρ denotes the Euclidean distance. It is well known (cf. Lemma A below) that there exists $\delta > 0$ (which depends only on Q and the dimension d) such that for all $x \in R^d$:

$$\frac{1}{2} \le E^x\{\, \exp(-QT(2\delta)) \,\}; \qquad E^x\{\, \exp(QT(2\delta)) \,\} \le 2. \tag{4}$$

In fact, the two expectations in (4) do not depend on x by the spatial homogeneity of the process. Now put

$$2r = \rho(K, \partial D) \wedge 2\delta. \tag{5}$$

Then for any $s < 2r$ we have, by the strong Markov property, since $T(s) < \tau_D$ under P^{x_0}:

(6) $$\infty > u(x_0) = E^{x_0}\{\ e(T(s))\ u(X(T(s)))\ \}$$

$$\geq E^{x_0}\{\ \exp(-QT(s))\ u(X(T(s))\ \}\ .$$

The isotropic property of the Brownian motion implies that the random variables $T(s)$ and $X(T(s))$ are stochastically independent for each s. Hence we obtain from (6) and the first inequality in (4):

(7) $$u(x_0) \geq \frac{1}{2} E^{x_0}\{\ u(X(T(s)))\ \}\ .$$

The expectation on the right side above is the area average of the values of u on the boundary of $B(x_0,s)$. Hence we obtain by integrating with respect to the radius:

(8) $$a_d \int_0^{2r} E^{x_0}\{\ u(X(T(s)))\ \}\ s^{d-1}\, ds = \int_{B(x_0,2r)} u(y)\, dy,$$

where $a_d\, s^{d-1}$ is the area of $\partial B(x_0,s)$. It follows from (7) and (8) that

(9) $$u(x_0) \geq \frac{1}{2V(2r)} \int_{B(x_0,2r)} u(y)\, dy,$$

where $V(2r)$ is the volume of $B(x_0,2r)$. [The terms "area" and "volume" used above have their obvious meanings in dimension $d = 1$ or 2.]

Next, let $x \in B(x_0,r)$ so that $\rho(x,\partial D) \geq r$ by (5). We have for $0 < s < r$:

(10) $$u(x) = E^{x}\{\ e(T(s))\ u(X(T(s)))\ \} \leq E^{x}\{\ \exp(QT(s))\ u(X(T(s)))\ \}$$

$$= E^{x}\{\ e(QT(s))\}\ E^{x}\{\ u(X(T(s)))\ \} \leq 2E^{x}\{\ u(X(T(s)))\ \}$$

by independence and the second inequality in (4). Integrating as before we obtain

$$u(x) \le \frac{2}{V(r)} \int_{B(x,r)} u(y)\, dy \quad . \tag{11}$$

Since $B(x,r) \subset B(x_0,2r)$ and $u \ge 0$, (9) and (10) together yield

$$u(x) \le 2^{d+2} u(x_0) \quad . \tag{12}$$

In particular we have proved that $u(x) < \infty$ if $\rho(x,x_0) < r$ and consequently we may interchange the roles of x_0 and x in the above. Since the number r is fixed independently of x, and K is compact, a familiar "chain argument" establishes the theorem. Indeed if N is the number of overlapping balls of fixed radius r which are needed to lead in a chain from any point to any other point in K, then the constant A in (3) may be taken to be $2^{(d+2)N}$. □

Corollary. If K is fixed and D is enlarged, the inequalities in (3) remain valid with the same constant A.

This is clear from the proof, and will be needed for the application in Theorem 3.1.

The following lemma plays a key role below. Its essential feature is that only the (Lebesgue) measure $m(E_n)$ of E_n, and not its shape or smoothness, is involved.

Lemma A. Let $\{E_n\}$ be sets with $m(E_n)$ decreasing to zero. Then we have for each $t > 0$:

(13) $$\lim_{n\to\infty} \sup_{x\in \bar{E}_n} P^x\{ \tau(E_n) > t \} = 0 .$$

For any constant Q we have

(14) $$\lim_{n\to\infty} \sup_{x\in \bar{E}_n} E^x\{ \exp(Q\tau(E_n)) \} = 1 .$$

Proof. We have for any E and $t > 0$:

(15) $$\sup_{x\in \bar{E}} P^x\{ \tau(E) > t \} \le \sup_{x\in \bar{E}} P^x\{ X(t) \in E \} \le \frac{m(E)}{(2\pi t)^{d/2}}$$

because the probability density of $X(t)$ is bounded by $(2\pi t)^{-d/2}$. This implies (13). Next we obtain from (15) followed by a Markovian iterative argument:

$$\sup_{x\in \bar{E}} P^x\{ \tau(E) > nt \} \le \left(\frac{m(E)}{(2\pi t)^{d/2}}\right)^n .$$

Therefore we have

$$E^x\{ \exp(Q(\tau(E))) \} \le \sum_{n=0}^{\infty} e^{Q(n+1)t}\, P^x\{ \tau(E) > nt \}$$

$$\le e^{Qt} \sum_{n=0}^{\infty} [\, e^{Qt}\, m(E)\, (2\pi t)^{-d/2} \,]^n .$$

Given Q, chose t so small that Qt is near zero. For this t, if $m(E)$ is small enough the infinite series above has a sum near 1. This proves (14). □

It follows from Theorem 1.1 that if $u \not\equiv \infty$ in D then $u < \infty$ in D. When $m(D) < \infty$, this result has a sharpening which is not valid in the usual analytical setting of Harnack inequalities, in which only local boundedness can be claimed. The situation will be clarified in later sections when we relate the function u to a positive solution of the Schrödinger equation.

Theorem 1.2. Let D be a domain with $m(D) < \infty$, and let q and f be as in Theorem 1.1, but f be bounded as well as nonnegative. If $u(q,f;\cdot) \not\equiv \infty$ in D, then it is bounded in $\bar{D}$.

Proof. Let us remark that if $m(D) < \infty$, then $P^x\{ \tau_D < \infty \} = 1$ for all $x \in R^d$, so that we may omit "$\tau_D < \infty$" in the definition (2). Write u as before and let $\|q\| = Q$. Let K be a compact subset of D such that $m(E) < \delta$ where $E = D-K$, and where δ is so small that

$$\sup_{x\in\bar{E}} E^x\{ \exp(Q\tau(E)) \} \leq 1+\varepsilon . \tag{16}$$

This is possible by Lemma A. Note that E is open and $\tau_E \leq \tau_D$. For $x \in \bar{E}$ let us put

$$\begin{aligned} u_1(x) &= E^x\{ e(\tau_D)\, f(X(\tau_D));\ \tau_E < \tau_D \}, \\ u_2(x) &= E^x\{ e(\tau_D)\, f(X(\tau_D));\ \tau_E = \tau_D \}. \end{aligned} \tag{17}$$

We have by the strong Markov property:

$$\begin{aligned} u_1(x) &= E^x\{ \tau_E < \tau_D;\ e(\tau_E)\, E^{X(\tau_E)}[e(\tau_D)\, f(X(\tau_D))] \} \\ &= E^x\{ \tau_E < \tau_D;\ e(\tau_E)\, u(X(\tau_E)) \}. \end{aligned} \tag{18}$$

On the set $\{ \tau_E < \tau_D \}$, we have $X(\tau_E) \in K$, and u is bounded on K by Theorem 1.1. Hence we have by (16) and (18):

(19) $$u_1(x) \le E^x\{ \exp(Q\tau_E) \} \|u\|_K \le (1+\varepsilon) \|u\|_K .$$

On the other hand, we have for $x \in \bar{E}$:

(20) $$u_2(x) \le E^x\{ e(\tau_E) f(X(\tau_E)) \}$$
$$\le E^x\{ e(\tau_E) \} \|f\| \le (1+\varepsilon) \|f\| .$$

Combining the last two inequalities we have

(21) $$u(x) \le (1+\varepsilon)(\|u\|_K + \|f\|).$$

Since $\bar{D}-\bar{E} \subset K$, (21) holds trivially for $x \in \bar{D}-\bar{E}$. Thus (21) holds for all $x \in \bar{D}$. □

It is clear how we can make more precise the dependence of ε in (21) on K, thereby giving an estimate of the global bound $\|u\|_{\bar{D}}$ in terms of a local bound $\|u\|_K$ and $\|f\|$. Theorem 1.2 is true without any condition on the smoothness of ∂D. In the probabilistic treatment of the Dirichlet problem a point z is said to be a regular boundary point iff $z \in \partial D$ and $P^z\{ \tau_D = 0 \} = 1$, namely iff z is regular for D^c. The equivalence of this definition of regularity with the classical definition based on the solvability of the boundary value problem is well known. The next result is an extension of the probabilistic solution to the Dirichlet problem (D,f) to the present setting when the Feynman-Kac functional e_q is attached to the Brownian motion process. It will be seen in §2 that this extension is tantamount to replacing the Laplacian operator Δ by the Schrödinger operator $\Delta+2q$. When $q \equiv 0$ the theorem below reduces to Dirichlet's first boundary value problem.

Theorem 1.3. Let D and q be as in Theorem 1.2, but $f \in b\mathcal{B}(\partial D)$. If z is a regular point of ∂D and f is continuous at z, then we have

$$\lim_{x \to z} u(x) = f(z). \tag{22}$$

Remark. Since u is defined in $\bar{D}$ it is natural that the variable x in (22) should vary in $\bar{D}$, and not just in D. This minor but nontrivial point is sometimes overlooked.

Proof. Without loss of generality we may suppose $f \geq 0$. Given $\varepsilon > 0$, there exists $\delta > 0$ such that

$$\sup_{x \in R^d} E^x\{ e^{2QT_r} \} \leq 1 + \varepsilon \quad \text{for } r \leq \delta; \tag{23}$$

$$\sup_{y \in B(z,2\delta) \cap (\partial D)} |f(y)-f(z)| \leq \varepsilon. \tag{24}$$

Let $x \in B(z,\delta)$, and $0 < r < \delta$. Write τ for τ_D and put

$$u_1(x) = E^x\{ T_r < \tau;\ e(\tau)\, f(X(\tau)) \},$$

$$u_2(x) = E^x\{ \tau \leq T_r;\ e(\tau)\, f(X(\tau)) \}.$$

It is well known that for each $t > 0$, $P^x\{ \tau > t \}$ is upper semicontinuous in $x \in R^d$. Since $P^z\{ \tau > t \} = 0$, it follows easily that

$$\lim_{x \to z} P^x\{ \tau > T_r \} = 0 \tag{25}$$

where $x \in \bar{D}$. We have by the strong Markov property:

$$u_1(x) = E^x\{ T_r < \tau;\ e(T_r)u(X(T_r)) \}.$$

Hence by Theorem 1.2 followed by Schwarz's inequality:

$$u_1(x) \le E^x\{ T_r < \tau;\ e^{QT_r} \} \|u\|_D$$

$$\le P^x\{ T_r < \tau \}^{\frac{1}{2}} E^x(e^{2QT_r})^{\frac{1}{2}} \| u\|_D .$$

Therefore $\lim_{x\to z} u_1(x) = 0$ by (23) and (25). Next we have by (24), since $X(\tau) \in B(z,2\delta)$ on $\{ \tau \le T_r \}$ under P^x:

$$|u_2(x)-E^x\{ \tau \le T_r;\ e(\tau)f(z) \}| \le E^x\{ \tau \le T_r;\ e^{QT_r} \}\varepsilon \le (1+\varepsilon)\varepsilon ;$$

and by (23):

$$|1-E^x\{ \tau \le T_r;\ e(\tau) \}| \le P^x\{ T_r < \tau \} + E^x\{ \tau \le T_r;\ e(\tau)-1 \}$$

$$\le P^x\{ T_r < \tau \} + E^x\{ e^{QT_r} \} - 1$$

$$\le P^x\{ T_r < \tau \} + \varepsilon.$$

Since ε is arbitrary, it follows from the above inequalities and (25) that $\lim_{x\to z} u_2(x) = f(z)$. Thus (22) is true. □

The intuitive content of Theorems 1.2 and 1.3 is this: the motion of the Brownian path in a domain is such that large positive values cancel large negative values of $q(X(t))$ so neatly that no after-effect is felt as it approaches the boundary, provided that cancellation is possible in an average sense, measured exponentially. Moreover the latter possibility is irrespective of the starting point of the path.

2. Schrödinger equation

Let D be a domain in R^d. We introduce the notation

$$(1) \qquad Q_t f(x) = E^x\{ t < \tau_D ; f(X_t) \}$$

for $f \in b\mathcal{B}$. Then $\{ Q_t, t \geq 0 \}$ is the transition semigroup of the Brownian motion killed upon the exit from D. Let

$$(2) \qquad G_D f(x) = E^x\{ \int_0^{\tau_D} f(X_t)\, dt \}$$

where the right member is defined first for $f \geq 0$, then through $f = f^+ - f^-$ in the usual way, provided either $G_D f^+$ or $G_D f^-$ is finite. We shall be concerned only with the case where $G_D|f| < \infty$. Let $C^{(0)}(D)$ and $C^{(k)}(D)$, $k \geq 1$, denote respectively the classes of continuous and k times continuously differentiable functions on D. We write $f \in H(D)$ and say that f is Hölder continuous in D, iff for any compact subset C of D there exist two constants $\alpha > 0$ and M such that $|f(x)-f(y)| \leq M|x-y|^\alpha$ for x and y in C. For a proof of the following lemma see e.g. [6; Chapter 4, §§5-6].

Lemma B. If f is locally bounded in D and $G_D|f| < \infty$ then $G_D f \in C^{(1)}(D)$. If in addition $f \in H(D)$ then $G_D f \in C^{(2)}(D)$, and

$$(3) \qquad \Delta(G_D f) = -2f .$$

On the other hand if $f \in C^{(2)}(D)$ then

$$G_D(\Delta f) = -2f + h,$$

where h is harmonic in D.

Let $q \in b\mathcal{B}$ as in §1. The Feynman-Kac semigroup $\{ K_t, t \geq 0 \}$ is defined as follows:

$$(4) \qquad K_t f(x) = E^x\{ e_q(t) f(X_t) \}$$

for $f \in b\mathcal{B}$. Actually Feynman considered a purely imaginary q and Kac a nonpositive q. For our q the semigroup need not be submarkovian. It is known that its infinitesimal generator is $\frac{\Delta}{2} + q$ (see [3]). When $q \equiv 0$, of course $\{K_t\}$ reduces to the Brownian semigroup $\{P_t\}$. In this case the function u in (2) of §1 is harmonic in D, namely it satisfies the Laplace equation $\Delta u = 0$ there. Theorem 1.1 becomes a classical Harnack theorem for harmonic functions and Theorem 1.3 becomes Dirichlet's first boundary value problem. We are now going to show that for a general bounded q the function u satisfies the Schrödinger equation (5) below.

Theorem 2.1. Let D, q and f be as in the definition (2) of §1 except that f need not be nonnegative. If $u(q, |f|; \cdot) \not\equiv \infty$ in D, then $u(q,f;\cdot) \in C^{(1)}(D)$. If in addition $q \in H(D)$, then $u(q,f;\cdot)$ satisfies the equation

$$(5) \qquad (\Delta+2q)u = 0 \qquad \text{in } D .$$

Proof. Since the conclusions are local properties let us begin by localization. Writing u as before we see that it is locally bounded by Theorem 1.1. Let B be a small ball such that $\bar{B} \subset D$, and

$$(6) \qquad \sup_{x \in \bar{B}} E^x\{ \exp(Q\tau_B) \} < \infty .$$

We have for $x \in B$:

$$u(x) = E^x\{ e(\tau_B)\, u(X(\tau_B)) \} . \tag{7}$$

Comparing this with the definition of u we see that we have replaced (D,f) by (B,u), where B is bounded and satisfies (6), and u is bounded in $\bar{B}$. We need only prove the conclusions of the theorem for x in B. Reverting to the original notation we may therefore suppose that the domain D has the properties of B above, in particular $\tau_D < \infty$ and $E^x\{ e(\tau_D) \}$ is bounded in $\bar{D}$; and f is bounded. These conditions will be needed in the use of Fubini's theorem in the calculations which follow. [Warning: one must check the finiteness of the quantities below when q and f are replaced by $|q|$ and $|f|$; the former replacement is *not* trivial.] We write τ for τ_D and put for $0 \le s < t$:

$$e(s,t) = \exp\left(\int_s^t q(X(r))\, dr \right),$$

thus $e(\tau) = e(0,\tau)$. We have by the Markov property:

$$E^x\{ 1_{\{s<\tau\}}\, e(s,\tau)\, f(X_\tau) | F_s \} = 1_{\{s<\tau\}}\, u(X_s). \tag{8}$$

This relation is used in the first and last equations below:

$$\begin{aligned} (9) \quad & E^x\{ \int_0^t 1_{\{s<\tau\}}\, q(X_s)\, u(X_s)\, ds \} \\ &= E^x\{ \int_0^{t\wedge\tau} q(X_s)\, e(s,\tau)\, f(X_\tau)\, ds \} \\ &= E^x\{ [e(\tau) - e(t \wedge \tau, \tau)]\, f(X_\tau) \} \end{aligned}$$

$$= E^x\{ t < \tau;\ [e(\tau)-e(t,\tau)]\ f(X_\tau)\} + E^x\{ t \ge \tau;\ [e(\tau)-1]\ f(X_\tau)\}$$

$$= E^x\{ e(\tau)\ f(X_\tau) \} - E^x\{ t < \tau;\ u(X_t) \} - E^x\{ t \ge \tau;\ f(X_\tau) \}.$$

Now put

$$v(x) = E^x\{ f(X(\tau_D)) \}$$

for $x \in D$. Then v is the probabilistic solution of the Dirichlet problem (D,f) reviewed above; hence $\Delta v = 0$ in D. The last member of (9) may be written as

$$u(x) - Q_t u(x) - v(x) + Q_t v(x).$$

Since both u and v are bounded, and $\lim_{t\to\infty} Q_t 1 = 0$ because D is bounded, we have $\lim_{t\to\infty} Q_t(u-v) = 0$. We may therefore let $t \to \infty$ in the first member of (9) to obtain, with the notation of (2):

(10) $$G_D(qu) = u-v \quad \text{in } D .$$

Since u as well as q is bounded, and $G_D 1 < \infty$ because D is bounded, we have $G_D(|qu|) < \infty$. Since v is harmonic it follows from (10) and Lemma B that $u \in C^{(1)}(D)$; if $q \in H(D)$ then $u \in C^{(2)}(D)$ and

$$\Delta u = \Delta v + \Delta G_D(qu) = -2qu$$

which is (5). □

Before going further let us recapitulate the essential part of Theorems 1.3 and 2.1, leaving aside the generalizations. Let D be a bounded domain, $q \in b\mathcal{B}(D) \cap H(D)$, $f \in C^{(0)}(\partial D)$. Suppose that for some x_0 in D we have $u(q,1;x_0) < \infty$, then writing $u(x)$ for $u(q,f;x)$, we have $u \in C^{(2)}(D)$ and u is a solution of the Schrödinger equation $(\Delta+2q)u = 0$ in D. Furthermore $u(x)$ converges to $f(z)$ as x approaches each regular point z of ∂D. In particular if ∂D is regular then $u \in C^{(0)}(\bar{D})$. For $q \equiv 0$, u is the well known solution to the Dirichlet problem (D,f). Now in the latter case there is a converse as follows. Let $\varphi \in C^{(2)}(D) \cap C^{(0)}(\bar{D})$ and $\Delta\varphi = 0$ in D, then we have for all x in D:

$$\varphi(x) = E^x\{ \varphi(X(\tau_D)) \} .$$

This provides an extension of Gauss's average theorem for harmonic functions and implies the uniqueness of the solution to the Dirichlet problem. We proceed to establish corresponding results in the present setting.

The following lemma is stated for the sake of explicitness.

Lemma C. Let D, q, f be as in the definition of (2) of §1, except that f need not be nonnegative. If $u(q,|f|;\cdot) \not\equiv \infty$ in D, then we have for all $x \in D$ and $t \geq 0$:

$$E^x\{ e(\tau_D)\, f(X(\tau_D)) | \mathcal{F}_t \} = e(t \wedge \tau_D)\, u(X(t \wedge \tau_D)) \quad . \tag{11}$$

Proof. We have

$$E^x\{1_{\{t<\tau_D\}} e(\tau_D) f(X(\tau_D)) | \mathcal{F}_t\} = 1_{\{t<\tau_D\}} e(t)\, E^x\{e(t,\tau_D) f(X(\tau_D)) | \mathcal{F}_t\}$$

$$= 1_{\{t<\tau_D\}}\, e(t)\, u(X_t)$$

by (8) (with s replaced by t). On the other hand,

$$E^x\{ 1_{\{t\geq\tau_D\}} e(\tau_D) f(X(\tau_D)) | \mathcal{F}_t \} = 1_{\{t\geq\tau_D\}} e(\tau_D) f(X(\tau_D))$$

because the trace of $\mathcal{F}_t$ on $\{ t \geq \tau_D \}$ contains the trace of $\mathcal{F}_{\tau_D}$ on $\{ t \geq \tau_D \}$. Now by Kellogg's theorem (see e.g. [6]) irregular points of ∂D form a polar set, hence $X(\tau_D)$ is a regular point of ∂D almost surely under P^x, $x \in D$; and consequently $u(X(\tau_D)) = f(X(\tau_D))$ by (2) of §1. Using this in the second relation above and adding it to the first relation we obtain (11). □

Theorem 2.2. Let D be an arbitrary domain and $q \in bC^{(0)}(D)$. Suppose that the function φ has the following properties:

$$\varphi \in C^{(2)}(D); \quad \varphi > 0 \text{ and } (\Delta+2q)\varphi = 0 \text{ in } D. \tag{12}$$

Then for any bounded subdomain E such that $\bar{E} \subset D$, we have

$$\forall\ x \in D: \ \varphi(x) = E^x\{ e(\tau_E)\, \varphi(X(\tau_E)) \} . \tag{13}$$

Proof. Although we can prove this result without stochastic integration, it is expedient to use Ito's formula. We have then in the customary notation:

$$d(e(t)\, \varphi(X_t)) = e(t)\{ \nabla\varphi(X_t)\, dX_t + (2^{-1}\Delta\varphi + q\varphi)(X_t)\, dt \} \tag{14}$$

where ∇ denotes the gradient and dX_t the stochastic differential (see e.g., [8]). The second term on the right side of (14) vanishes for $t < \tau_D$ by (12), and the first term is a local martingale. Since

$\bar{E}$ is compact and $\varphi \in C^{(2)}$ in a neighborhood of $\bar{E}$, we have

$$(15) \qquad \sup_{0 \le s \le t\wedge\tau_E} \| e(s) \nabla\varphi(X_s) \| \le e^{\|q\| t} \| \nabla\varphi \|_{\bar{E}} < \infty.$$

Hence (14) has the formal expression:

$$(16) \quad e(t\wedge\tau_E)\varphi(X(t\wedge\tau_E)) - e(0)\varphi(X(0)) = \int_0^{t\wedge\tau_E} e(s) \nabla\varphi(X_s)\, dX_s = M_t,$$

say, where $\{ M_t, F_t, t \ge 0 \}$ is a martingale, under P^x for each $x \in D$, with $E^x(M_t) = 0$ for $t \ge 0$. Taking E^x in (16) we obtain

$$(17) \qquad \varphi(x) = E^x\{ e(t\wedge\tau_E)\, \varphi(X(t\wedge\tau_E)) \} .$$

By enlarging E if necessary, we may assume that all boundary points of E are regular. By Theorem 1.3, the function v defined by

$$(18) \qquad v(x) = E^x\{ e(\tau_E)\, \varphi(X(\tau_E)) \}$$

is continuous on $\bar{E}$ and equals φ on ∂E. We can apply Lemma C with D and f replaced by E and φ to obtain

$$(19) \qquad E^x\{ e(\tau_E)\, \varphi(X(\tau_E)) \mid F_t \} = e(t\wedge\tau_E)\, v(X(t\wedge\tau_E)).$$

Since $\varphi > 0$, v cannot vanish; and since v is continuous, it is bounded below on $\bar{E}$. We conclude from (19) that

$$(20) \qquad e(t\wedge\tau_E) \le (\inf_{\bar{E}} v)^{-1} E^x\{ e(\tau_E)\, \varphi(X(\tau_E)) \mid F_t \} .$$

A simple consequence of (2) is that the family $e(t\wedge\tau_E)$, $t \ge 0$, of random variables is uniformly integrable. We can thus let t tend to infinity in (17) to obtain

$$\varphi(x) = E^x\{ e(\tau_E)\, \varphi(X(\tau_E)) \}, \tag{21}$$

which shows (13) for $x \in \bar{E}$. Note that (13) is trivial for $x \in D - \bar{E}$.

Let us introduce the notation, for any $E \subset D$:

$$u_E(x) = E^x\{ e(\tau_E) \} ; \tag{22}$$

in particular u_D is $u(q,1;\cdot)$ in our previous notation. Parts of Theorem 2.2 are important enough to be stated as corollaries.

Corollary 1. For each bounded domain (or open set) E such that $\bar{E} \subset D$, we have u_E bounded in D.

Corollary 2. If there exists a function φ satisfying the conditions in (12) and is furthermore bounded above and bounded away from zero, then u_D is bounded in $\bar{D}$. In particular, this is the case if D is bounded and $\varphi \in C^{(0)}(\bar{D})$.

Proof. For then we have from (13), for all $x \in D$:

$$(\inf_{\bar{D}} \varphi)\, u_E(x) \le \varphi(x) \le \sup_{\bar{D}} \varphi$$

and consequently u_E is uniformly bounded with respect to E. There exists a sequence of such subdomains E_n increasing to D so that

$\tau(E_n)$ increases to $\tau(D)$. Hence it follows by Fatou's lemma that

$$E^x\{\, e(\tau_D) \,\} \le \underline{\lim}_n \, E^x\{\, e(\tau(E_n)) \,\}$$

is also bounded in D. Now it is easily shown that u_D is indeed bounded in $\bar{D}$. □

When $q \ge 0$, $\bar{D}$ is compact and ∂D is regular, the second assertion in Corollary 2 was proved by Khas'minskii [4] for strong Markov processes with strong Feller property and continuous paths. His method uses the Taylor series for $e(\tau_D)$, an iterative argument à la Picard, and a maximum principle. These depend essentially on the non-negativeness of q. The methods used in this paper can also be partially generalized to the class of processes considered by him, but the more difficult theorems such as Theorem 1.2 elude us at the moment.

Without the further assumption in Corollary 2 we cannot conclude that $u_D \not\equiv \infty$ in D; nor that (13) holds when E is replaced by D. The simplest example is in R^1 when $D = (0,\pi)$, $q = 1/2$, $\varphi(x) = \sin x$. The fact that $u_D \equiv \infty$ in this example will follow by contraposition from the results below. We need an easy but important lemma.

Lemma D. Let D be a domain with $m(D) < \infty$, and $u_D \not\equiv \infty$ in D. There exists a constant $c_0 > 0$ such that for all $x \in \bar{D}$:

$$u_D(x) \ge c_0(1 \vee u_E(x)) \tag{23}$$

where E is any set with $\bar{E} \subset D$. For all D and q such that $m(D) \le M < \infty$ and $\|q\| \le Q < \infty$, the constant c_0 depends only on M and Q.

Proof. It follows from (15) of §1 that for some t_0 depending only on M we have $P^x\{\tau_D \le t_0\} \ge 1/2$ for all $x \in \bar{D}$. We have then

$$u_D(x) \ge E^x\{ e(\tau_D); \tau_D \le t_0 \} \ge E^x\{ e^{-Qt_0}; \tau_D \le t_0 \} \ge c_0,$$

where $c_0 = e^{-Qt_0}/2$. If $\bar{E} \subset D$ we have $\tau_E < \tau_D$ and so by the strong Markov property

$$u_D(x) = E^x\{ e(\tau_E)\, u_D(X(\tau_E)) \} \ge E^x\{ e(\tau_E) \}\, c_0 = u_E(x)\, c_0. \qquad \square$$

A simple consequence of Lemma D is the following converse to Corollary 2 above. If $m(D) < \infty$, and $u_D = \varphi$, then this φ satisfies (12); it is bounded above by Theorem 1.2 and bounded away from zero by Lemma D. In particular if D is bounded and ∂D is regular then $u_D \in C^{(0)}(\bar{D})$ by Theorem 1.3. The importance of this particular solution of the Schrödinger equation will become apparent in what follows.

Theorem 2.3. Let D be a domain with $m(D) < \infty$ and $u_D \not\equiv \infty$. Suppose that the function φ has the following properties:

$$(24) \qquad \varphi \in C^{(2)}(D) \cap bC^{(0)}(\bar{D}); \qquad (\Delta + 2q)\varphi = 0 \text{ in } D.$$

Then we have

$$(25) \qquad \forall\; x \in D: \qquad \varphi(x) = E^x\{ e(\tau_D)\, \varphi(X(\tau_D)) \}.$$

Proof. Although φ is no longer positive, the proof of Theorem 2.2 needs no change up to (17) there. Now by (23), $u_E \le c_0^{-1}\, u_D < \infty$ in D; hence $e(\tau_E)$ is integrable under P^x, $x \in \bar{E}$. We may therefore apply Lemma C to E with $f \equiv 1$ to obtain

$$(26)\quad E^x\{\, e(\tau_E) | F_t \,\} = e(t \wedge \tau_E)\, u_E(X(t \wedge \tau_E)) \geq e(t \wedge \tau_E)\, c_0$$

since $\inf_{\bar{E}} u_E \geq c_0$ by Lemma D. It follows that the family of random variables $\{\, e(t \wedge \tau_E),\ t \geq 0 \,\}$ is uniformly integrable. Since φ is bounded continuous in $\bar{D}$ we may let $t \to \infty$ under E^x in (17) to deduce

$$(27)\qquad \varphi(x) = E^x\{\, e(\tau_E)\, \varphi(X(\tau_E)) \,\}, \qquad x \in D.$$

Next we have, since $u_D < \infty$:

$$(28)\qquad E^x\{\, e(\tau_D) | F(\tau_E) \,\} = e(\tau_E)\, u_D(X(\tau_E)) \geq e(\tau_E)\, c_0 \quad ,$$

by Lemma D. Hence the family $\{e(\tau_E)\}$ is uniformly integrable as E ranges over all sets with closures contained in D. Let E_n be open bounded, $\bar{E}_n \subset E_{n+1} \subset D$ and $\bigcup_n E_n = D$. We obtain (25) by putting $E = E_n$ in (27) and passing to the limit as before.

Let us remark that under the assumptions of Theorem 2.3 it is possible to apply Ito's formula directly to D instead of E as we did, provided that we can deduce the boundedness of $\nabla\varphi$ in (14) from that of φ (and hence of $\Delta\varphi$). This would require some kind of global estimate of Schauder's type which might require stronger smoothness conditions.

If $q \equiv 0$ and D is bounded, the preceding theorem reduces to the classical result of the representation of a harmonic function by its boundary values. We are now ready to solve a similar problem for the Schrödinger equation.

Theorem 2.4. Let D be as in Theorem 2.3, ∂D be regular, and $q \in H(D)$. For any $f \in bC^0(\bar{D})$ there is a unique φ satisfying (24) such that $\varphi \equiv f$ on ∂D. Indeed this φ is given by

$$\varphi(x) = E^x\{\, e(\tau_D)\, f(X(\tau_D))\,\}, \qquad x \in D. \tag{29}$$

Proof. That this φ satisfies (24) is proved by Theorem 2.1 and Theorem 1.3. The uniqueness follows from Theorem 2.3. □

The relationship between the preceding results and those obtainable by the usual methods of partial differential equations will be discussed in a separate publication.

3. Further results

As a by-product of the methods used above which can be stated without mentioning probability, we give the following theorem about positive solutions of the Schrödinger equation. This will be needed in the next theorem.

Theorem 3.1. Let D be a domain and D_n be bounded domains such that $\bar{D}_n \subset D_{n+1}$ and $\bigcup_n D_n = D$. Let $q_n \in H(D_n)$ and $q_n \to q$ boundedly where $q \in H(D)$. Suppose that for each $n \geq 1$, there exists φ_n such that

$$\varphi_n > 0, \qquad (\Delta + 2q_n)\varphi_n = 0 \quad \text{in } D_n. \tag{1}$$

Then there exists φ such that

(2) $\varphi > 0, \quad (\Delta+2q)\varphi = 0 \quad \text{in } D.$

Proof. Let Q be a common bound of all $\| q_n \|$ (and $\| q \|$). By Corollary 1 to Theorem 2.2, the existence of φ_n implies that for each $n \geq 2$ the function u_n defined by

(3) $$u_n(x) = E^x\{ e_{q_{n+1}}(\tau_{D_n}) \}$$

is bounded in $\bar{D}_n$. Choose any $x_0 \in D_1$ and put

(4) $$v_n(x) = u_n(x)/u_n(x_0) \ .$$

According to the Corollary to Theorem 1.1, applied for $k > n$ to v_k on $\bar{D}_n$ and with $f \equiv 1$, there exists a constant $A_n > 0$ (depending only on D_{n+1}, D_n and Q) such that we have for all $k > n$:

(5) $$A_n^{-1} \leq \inf_{\bar{D}_n} v_k \leq \sup_{\bar{D}_n} v_k \leq A_n.$$

Define the measures below:

(6) $$\mu_k^{\pm}(dx) = q_{k+1}^{\pm}(x)v_k(x)dx, \quad \mu_k(dx) = \mu_k^{+}(dx)-\mu_k^{-}(dx) = q_{k+1}(x)v_k(x)dx$$

where $q_k = q_k^{+} - q_k^{-}$ is the usual decomposition. The two sequences of measures $\{ \mu_k^{\pm}, k > n \}$ on D_n are vaguely bounded because

(7) $$\mu_k^{\pm}(D_n) \leq QA_n \, m(D_n), \qquad k > n.$$

We have by Theorem 2.1: $v_k \in C^{(2)}(D_k)$ and

(8) $$(\Delta+2q_{k+1})v_k = 0, \qquad \text{in } D_k .$$

Hence it follows from Lemma B that for $k > n$:

(9) $$v_k = G_{D_n}(q_{k+1}\, v_k) + h_k$$

where h_k is harmonic in D_n. We have by (9) and (5):

(10) $$\| h_n \|_{D_n} \le A_n + QA_n\, G_{D_n} 1 < \infty .$$

Hence by Harnack's theorem on harmonic functions, followed by a diagonal argument, there exists a sequence $\{k_j\}$ such that $\{h_{k_j}\}$ converges uniformly on each D_n, $n \ge 1$; and the limit is a harmonic function h in D. By (7), the sequence $\{k_j\}$ may be chosen so that both $\mu^+_{k_j}$ and $\mu^-_{k_j}$ converge vaguely on each D_n, $n \ge 1$. The bounds in (5) then imply that for every $f \in L^1(D)$ (not only for $f \in bC(D_n)$), $\int_{D_n} f d\mu_{k_j}$ converges as $j \to \infty$. Since D_n is bounded, we know that $G_{D_n}(x,dy) = g_{D_n}(x,y)\,dy$ where $g_{D_n}(x,\cdot) \in L^1(D_n)$. Here the function g_{D_n} is the Green's function for D_n. Therefore if we substitute k_j for k in (9), the first term on the right side converges as $j \to \infty$. Thus $\lim_j v_{k_j} = v$ exists on D_n for each $n \ge 1$, and we obtain from (9):

(11) $$v = G_{D_n}(qv) + h \qquad \text{in } D_n.$$

Since $v \ge A_n^{-1}$ in D_n by (5), $v > 0$ in D. Using Lemma C, we see first that $v \in C^{(1)}(D)$ since qv is bounded by (5), and then $v \in C^{(2)}(D)$ since $qv \in H(D)$; finally by (11):

(12) $$\Delta v = \Delta G_{D_n}(qv) + \Delta h = -2qv.$$

This is true in D_n for $n \geq 1$, hence also in D. We have proved the existence of φ in (2) since v is such a function. □

Several experts in partial differential equations were consulted about Theorem 3.1. Hans Weinberger said it could be proved by classical eigenvalue methods of solving the Dirichlet problem for strongly elliptic equations. S.T. Yau said it could be proved (and the bounded convergence of q_n generalized) by variational methods. N. Trudinger said he would prove it by using Harnack's inequalities as we did above. A related result is given in [5].

From our point of view, the interest of Theorem 3.1 lies in the observation that its probabilistic analogue is false. Namely, if D_n increases to D, the finiteness of u_{D_n} for all n does not imply the finiteness of u_D. Yet u_{D_n} satisfies (1) (with $q_n \equiv q$) for each n, and if $u < \infty$ it will satisfy (2). A counterexample is furnished by the example in R^1 given above.

Let

$$L_t f(x) = E^x\{\ t < \tau_D;\ e(t)f(X_t)\ \}$$

in analogy with (4) of §2. In the next theorem we relate the finiteness of u_D to several conditions on the semigroup $\{L_t\}$ just defined. The results hold for the u in (2) of §1 with bounded f, but we put $f \equiv 1$ for simplicity. Since D and q are fixed below we will write u for u_D, τ for τ_D and $e(t)$ for $e_q(t)$. Consider then the following statements:

(a) $u \not\equiv \infty$ in D;

(b) for every $x \in D$, we have

(13) $$\int_0^\infty L_t 1(x)dt < \infty ;$$

(c) there is at least one x_0 in D having the property that for every $\delta > 0$, there exist $A(\delta)$ and $N(\delta)$ such that

(14) $$L_t\, 1(x_0) \le A(\delta)\, e^{\delta t} \qquad \text{for } t \ge N(\delta) \;;$$

(d) there exists φ satisfying (12) of §2 above;

(e) for any bounded open set E such that $\bar{E} \subset D$, u_E is bounded in $\bar{E}$.

Theorem 3.2. For any domain D, (a) and (b) are equivalent; and (b) implies (c). If $q \in bC^{(0)}(D)$, then (d) implies (e). If $q \in H(D)$, then (c) implies (d).

Proof. By Theorem 1.1, (a) implies that $u(x) < \infty$ for all $x \in D$. For any $s > 0$ and $n \ge 1$ we have by the Markov property:

(15) $$E^x\{e(\tau);\ ns < \tau \le (n+1)s\} = E^x\{e(ns);\ ns < \tau;\ E^{X(ns)}[e(\tau);\ 0<\tau\le s]\}.$$

Using (15) of §1, we can choose s so that $C = \inf_{x\in R^d} P^x\{\tau \le s\} > 0$. Then it is trivial that for every $y \in D$, we have

(16) $$Ce^{-Qs} \le E^y\{\ e(\tau);\ 0 < \tau \le s\ \} \le e^{Qs}$$

where $Q = \|q\|$. It follows from (15) and (16) that (a) is equivalent to:

(17) $$\sum_{n=0}^{\infty} E^x\{\ e(ns);\ ns < \tau\ \} < \infty.$$

For $ns < t < (n+1)s$, we have $e^{-Qs}e(ns) \le e(t) \le e^{Qs}e(ns)$. Hence (17) is equivalent to (13) by an easy comparison. This proves that (a) and (b) are equivalent. Of course, (c) is a much weakened form of (b) via (17).

Next, for a fixed $\varepsilon > 0$, we have by (15) and the second inequality in (16) applied to the functional $e_{q-\varepsilon}$ instead of e_q:

$$E^x\{ e_{q-\varepsilon}(\tau);\ n < \tau \le n+1 \} \le e^{Q+\varepsilon}\ E^x\{ e_{q-\varepsilon}(n);\ n < \tau \}$$

$$= e^{Q+\varepsilon-n\varepsilon}\ E^x\{ e_q(n);\ n < \tau \} \quad .$$

If (c) is true, the last member above is bounded by $e^{Q+\varepsilon+n(\delta-\varepsilon)}\ A(\delta)$ for $n \ge N(\delta)$. We may choose $0 < \delta < \varepsilon$; then

$$(18) \qquad E^x\{e_{q-\varepsilon}(\tau)\} = \sum_{n=0}^{\infty} E^x\{e_{q-\varepsilon}(\tau);\ n<\tau\le n+1\} \le A(\delta)\ e^{Q+\varepsilon} \sum_{n=0}^{\infty} e^{n(\delta-\varepsilon)} < \infty .$$

Now assume $q \in H(D)$. We can then apply Theorem 2.1 to obtain $u_{q-\varepsilon} \in C^{(2)}(D)$ and

$$(19) \qquad (\Delta+2(q-\varepsilon))\ u_{q-\varepsilon} = 0.$$

Since this is true for every $\varepsilon > 0$, and $q-\varepsilon$ converges to q boundedly as $\varepsilon \to 0$, we can apply Theorem 3.1 to conclude that (d) is true. Finally, if $q \in bC^{(0)}(D)$, then (d) implies (e) by Corollary 1 to Theorem 2.2.

The conditions in (b) and (c) are meaningful in the spectral theory of the semigroup $\{L_t\}$, or its infinitesimal generator the Schrödinger operator. At least when (14) holds for all x in D, this interpretation should yield (d) in some sense as a result on the point spectrum of the operator. It is not clear under what precise conditions the classical approach will confirm the results above obtained by probabilistic methods.

References

[1] K.L. CHUNG. On stopped Feynman-Kac functionals. *Séminaire de Probabilités XIV (Univ. Strasbourg).* Lecture Notes in Math. *784*, Springer-Verlag, Berlin, 1980.

[2] K.L. CHUNG and K.M. RAO. Sur la theorie du potentiel avec la fonctionelle de Feynman-Kac. *C.R. Acad. Sci. Paris 210* (1980).

[3] M. KAC. On some connections between probability theory and differential and integral equations. *Proc. Second Berkeley Symposium on Math. Stat. and Probability*, 189-215. University of California Press, Berkeley, 1951.

[4] R.Z. KHAS'MINSKII. On positive solutions of the equation Au + Vu = 0. *Theo. Prob. Appl. 4* (1959), 309-318.

[5] W.F. MOSS and J. PIEPENBRINK. Positive solutions of elliptic equations. *Pac. J. Math. 75* (1978), 219-226.

[6] S. PORT and C. STONE. *Brownian Motion and Classical Potential Theory.* Academic Press, New York, 1978.

[7] B. SIMON. *Functional Integration and Quantum Physics.* Academic Press, New York, 1979.

[8] D.W. STROOCK and S.R.S. VARADHAN. *Multidimensional Diffusion Processes.* Springer-Verlag, Berlin, 1979.

[9] E.B. DYNKIN. *Markov Processes.* Springer-Verlag, Berlin, 1965.

K.L. Chung
Department of Mathematics
Stanford University
Stanford, CA 94305, U.S.A.

K.M. Rao
Department of Mathematics
University of Florida
Gainesville, FL 32611, U.S.A.

TWO RESULTS ON DUAL EXCURSIONS

by

R.K. GETOOR and M.J. SHARPE

1. Introduction

This paper is a sequel to the recent work [5]. As in that paper, it is supposed for the two principal results that one is given a pair $X,\hat{X}$ of standard Markov processes on a common state space E having a dual density relative to some σ-finite measure on E. This condition is considerably stronger than the usual duality of resolvents. In fact, it was shown in [5] that duality of densities is equivalent to classical duality of certain space-time processes associated with $X,\hat{X}$. The main result of [5] concerned the construction and properties of families of measures $P^{x,\ell,y}$ which were shown to govern the distribution of excursions of X from a given closed homogeneous optional set M, conditional on the excursion starting at x, ending at y and having length ℓ. The precise meaning of "govern" in the statement above was laid out in two situations in [5]. One considers the excursion straddling a stopping time T. In the first case, one considers the stopping times defined by Maisonneuve [8]. See §3 for a formal description. Roughly speaking, such stopping times correspond to rules where the decision to stop during an excursion may be based only on

*Research supported in part by NSF Grant MCS79-23922.

information available before the excursion began, and on the age process of the excursion. In this case it was shown that the excursion straddling T has conditional laws precisely the $P^{x,\ell,y}$ described above. The specific form of T is of no importance, all of its influence having been absorbed by conditioning on the length ℓ of the excursion. In the second case, it was shown [5,§13] that if T is a terminal time, then the excursion straddling T has conditional laws $P^{x,\ell,y}[\ \cdot \mid T < \zeta\ ]$.

It was not until the authors heard Pitman's lecture on his paper [10] that they were able to proceed with the case of the excursion straddling a general stopping time T. One represents a general stopping time T in the form $T(\omega) = H(t,\omega;\theta_t\omega)$, where H is predictable in (t,ω) and $\omega' \to (H(t,\omega;\omega')-t)^+$ is a stopping time for every t, ω. This representation is proved in §2. Let $G(\omega)$ denote the left endpoint of the excursion straddling $T(\omega)$, and set $A_\omega(\omega') = [H(G(\omega),\omega;\omega')-G(\omega)]^+$. The main result of §3 states in essence that the conditional law of the excursion straddling T, given all information outside the excursion interval, its length L and its endpoint $X(D-)$, is $P^{X(G),L,X(D-)}[\ \cdot \mid A_\omega < \zeta\ ]$. The idea of the above representation of T comes from Pitman, though there are technical differences between our presentations*. In addition, a key step (3.9) in our proof is really due to Pitman. This result does not require duality and Pitman does not assume duality hypotheses. These hypotheses enter only when considering the $P^{x,\ell,y}$ and our main result (3.5) may be thought of as a "disintegration" of Pitman's under these stronger hypotheses.

---

*In fact, Pitman has pointed out that the result of §2 is a sharpening of Theorem 5.3 of Courrége and Priouret "Temps d'arrêt d'une fonction aléatoire: relations d'équivalence associées et propriétés de décomposition, "*Publ. Inst. Statist. Univ. Paris* 14 (1965), 245-274.

The second result of this paper occupies §4. We show that if b is a regular point for X, if b is recurrent, and if all excursions of X from b begin and end at b, then the path map Φ on Ω that reverses every excursion away from b maps P^b to $\hat{P}^b$. This result is related to a similar result proved in [5] for self dual processes.

2. A Representation Theorem for Stopping Times

In this section we suppose that X is a right process with state space $(E,\mathcal{E})$ as described in [2] or [11]. In fact, it suffices to assume that X is a right continuous strong Markov process with a U-space E as state space. In the next section we shall use the results of this section only when X is a standard process, and so the reader may suppose this from the beginning if he prefers.

We introduce some notation following [11]. The basic filtrations are defined by,

$$\mathcal{F}_t^{oo} \equiv \sigma\{ f(X_s): \ s \le t, \ f \in b\mathcal{E} \}; \qquad \mathcal{F}^{oo} \equiv \sigma(\bigcup_{t\ge 0} \mathcal{F}_t^{oo})$$

$$\mathcal{F}_t^{o*} \equiv \sigma\{ f(X_s): \ s \le t, \ f \in b\mathcal{E}^*\}; \qquad \mathcal{F}^{o*} \equiv \sigma(\bigcup_{t\ge 0} \mathcal{F}_t^{o*})$$

where $\mathcal{E}^*$ is the σ-algebra of universally measurable sets over $(E,\mathcal{E})$, and "≡" means "defined to be equal to". Clearly, $\mathcal{F}_t^{oo} \subset \mathcal{F}_t^{o*}$ for each $t \ge 0$. If μ is a probability on $(E,\mathcal{E}^*)$, $\mathcal{F}_t^{\mu} \equiv \mathcal{F}_t^{o*} \vee \mathcal{N}^{\mu}$ where $\mathcal{N}^{\mu}$ consists of all P^{μ} null sets in the P^{μ} completion, $\mathcal{F}^{\mu}$, of $\mathcal{F}^{o*}$. Finally $\mathcal{F}_t \equiv \bigcap_{\mu} \mathcal{F}_t^{\mu}$, $\mathcal{F} \equiv \bigcap_{\mu} \mathcal{F}^{\mu}$, $\mathcal{N} \equiv \bigcap_{\mu} \mathcal{N}^{\mu}$ where the intersections are over all probabilities μ on $(E,\mathcal{E}^*)$. It is shown in (3.9) of [11] that $\mathcal{F}^{\mu}$ is also the P^{μ} completion of $\mathcal{F}^{oo}$ and that $\mathcal{F}_t^{\mu} = \mathcal{F}_t^{oo} \vee \mathcal{N}^{\mu}$. We shall need the basic fact

(2.1) $$\mathcal{F}_t = \mathcal{F}_0^{o*} \vee \mathcal{F}_t^{oo} \vee \mathcal{N}$$

which is proved in (3.10) of [11]. Let $\mathcal{I}$ be the class of extended real valued evanescent processes; that is, $Z = (Z_t(\omega))$ is in $\mathcal{I}$ if

$$\{\ \omega:\ \exists\ t \geq 0 \ \text{ with } Z_t(\omega) \neq 0\ \} \in \mathcal{N}\ .$$

Let $\mathcal{O}^*$ be the optional σ-algebra over $(\mathcal{F}_t^{o*})$, and $\mathcal{P}^*$ the predictable; $\mathcal{O}^*$ (respectively, $\mathcal{P}^*$) is generated by $\mathcal{I}$ and the processes Z that are adapted to $(\mathcal{F}_t^{o*})$ and are such that for each $\omega \in \Omega$, $t \to Z_t(\omega)$ is right continuous on $[0,\infty[$ and has left limits on $]0,\infty[$ (respectively, $t \to Z_t(\omega)$ is left continuous on $]0,\infty[$). Let $\mathcal{O}$ and $\mathcal{P}$ be the optional and predictable σ-algebras over $(\mathcal{F}_{t+})$ respectively; these are defined by replacing $(\mathcal{F}_t^{o*})$ with $(\mathcal{F}_{t+})$ in the previous sentence. Obviously $\mathcal{O}^* \subset \mathcal{O}$ and $\mathcal{P}^* \subset \mathcal{P}$. It is well-known and easy to see that $\mathcal{P}^* \subset \mathcal{O}^*$ and $\mathcal{P} \subset \mathcal{O}$. (Processes of the form $t \to 1_\Lambda 1_{]s,\infty[}(t)$ with $\Lambda \in \mathcal{F}_s^{o*}$ generate $\mathcal{P}^*$, and those with $\Lambda \in \mathcal{F}_{s+}$ generate $\mathcal{P}$.) If X is a right process, then $\mathcal{F}_{t+} = \mathcal{F}_t$ for all t, and so, $\mathcal{P}$ and $\mathcal{O}$ are the usual predictable and optional σ-algebras over X in this case. See [11].

(2.2) THEOREM. *Let* T *be a* $(\mathcal{F}_{t+})$ *stopping time. Let* $G_t \equiv \mathcal{F}^{o*} \vee \mathcal{F}_t^{oo} \subset \mathcal{F}_t^{o*}$ *and* $G \equiv \sigma(\bigcup_{t\geq 0} G_t) \subset \mathcal{F}^{o*}$. *Then there exist* $H \in \mathcal{P}^* \times \mathcal{F}^{o*}$ *and a* (G_{t+}) *stopping time* T^* *with* $T = T^*$ *almost surely (i.e.* $\{T \neq T^*\} \in \mathcal{N}$*) such that*

(2.3) $T^*(\omega) = H(t,\omega;\theta_t\omega)$ *on* $\mathbb{R}^+ \times \Omega$,

(2.4) $\omega' \to (H(t,\omega;\omega')-t)^+$ *is a* (G_{t+}) *stopping time for* $(t,\omega) \in \mathbb{R}^+ \times \Omega$.

PROOF. Let T be a (F_{t+}) stopping time taking on only the values a and ∞ with $\{T = a\} \in F_a$. By the usual dyadic approximation any (F_{t+}) stopping time is a countable infimum of such two-valued stopping times; and since (2.3) and (2.4) are preserved under the operation of taking a countable infimum, it suffices to prove the theorem for such T. By (2.1), there exists $\Lambda \in F_0^{o*} \vee F_a^{oo}$ with $\{T = a\} = \Lambda$ almost surely. Then $T^* = a$ on Λ and $T^* = \infty$ on Λ^c is a (G_t) stopping time and $T = T^*$ almost surely. Suppose first that

$$(2.5) \qquad \Lambda = \{ (X_{t_0}, X_{t_1}, \ldots, X_{t_n}) \in A \}$$

where $0 = t_0 < t_1 < \cdots < t_n \leq a$ and $A \in E^* \times E^n$ where E^n is the product Borel field in the n-fold product E^n of E. Note that E^n is the usual product σ-algebra of E^n since E is separable and metrizable. Define $\varphi_A(x) = 1$ if $x \in A$, $\varphi_A(x) = \infty$ if $x \notin A$; that is, $\varphi_A = (1_A)^{-1}$. For Λ of the form (2.5) define

$$H(t,\omega;\omega') = a[\, 1_{\{0\}}(t)\varphi_\Lambda(\omega') +$$

$$\sum_{j=0}^{n-1} 1_{]t_j,t_{j+1}]}(t)\, \varphi_A(X_{t_0}(\omega),\ldots,X_{t_j}(\omega),X_{t_{j+1}-t}(\omega'),\ldots,X_{t_n-t}(\omega'))$$

$$+ \varphi_\Lambda(\omega)\, 1_{]t_n,\infty[}(t)\,].$$

Since $(x_{j+1},\ldots,x_n) \to \varphi_A(x_0,x_1,\ldots,x_n)$ is Borel measurable (E^{n-j}) for all $j \geq 0$ and $s \to X_s(\omega')$ is Borel (even right continuous) for all ω', it is evident that $H \in P^* \times G$ and that (2.3) holds. If $t_j < t \leq t_{j+1}$ and ω is fixed, $H(t,\omega;\omega')$ takes the value a on the ω' set

$$\Gamma = \{ \omega' : \varphi_A(X_{t_0}(\omega),\ldots,X_{t_j}(\omega),X_{t_{j+1}-t}(\omega'),\ldots,X_{t_n-t}(\omega')) = 1 \}$$

which is in F^{oo}_{a-t} and ∞ on Γ^c. Thus $\omega' \to H(t,\omega;\omega')-t$ is a (F^{oo}_t) stopping time if $0 < t \le t_n$. If $t = 0$, $H(0,\omega;\omega') = a\varphi_\Lambda(\omega') = T^*(\omega')$ which is a (G_{t+}) stopping time. If $t > t_n$, $(H(t,\omega;\omega')-t)^+$ is constant in ω' taking either the value $(a-t)^+$ or ∞ according as $\omega \in \Lambda$ or not. Sets Λ of the form (2.5) form an algebra generating $F^{o*}_0 \vee F^{oo}_t$ and since the $P^* \times F^{o*}$ measurability of H, (2.3), and (2.4) are preserved under monotone limits, Theorem 2.2 is established for all T of this special form. This completes the proof of (2.2).

(2.6) REMARK. Note that the proof shows that if $t > 0$ one may suppose that $\omega' \to (H(t,\omega;\omega')-t)^+$ is an (F^{oo}_{t+}) stopping time, and that the restriction of $H(t,\omega;\omega')$ to $]0,\infty[\times \Omega \times \Omega$ is in $P^* \times F^{oo}$ restricted to this set. From many points of view a predictable σ-algebra, for example P^*, should be considered as a σ-algebra on $]0,\infty[\times \Omega$ rather than $[0,\infty[\times \Omega$.

3. Excursion Straddling a Stopping Time

In this section we adopt the hypotheses of [5]. In particular X and $\hat{X}$ are standard processes with a Lusinian state space $(E,\mathcal{E})$ having a dual density $p(t,x,y)$ relative to a σ-finite measure $\xi(dx) = dx$ on E. This means that $P^x[X_t \in dy] = p(t,x,y)dy$, $\hat{P}^x[X_t \in dy] = p(t,y,x)dy$, and that p satisfies the Chapman-Kolmogorov equation for densities identically. We take Ω to be the space of paths ω: $\mathbb{R}^+ \to E \cup \{\delta\}$ which are right continuous on $[0,\infty[$ and have left limits in E on $]0,\zeta[$. As usual $\zeta(\omega) = \inf\{ t: \omega(t)=\delta \}$ where δ is a cemetery point adjoined to E as an isolated point. Then

$X_t(\omega) = \hat{X}_t(\omega) = \omega(t)$, and it is the families (P^x) and $(\hat{P}^x)$ of measures on (Ω, F^o) which describe X and $\hat{X}$. (In this section F^o and F^o_t are the σ-algebras that were denoted by F^{oo} and F^{oo}_t in §2. The change in notation is natural since in this section we are dealing with standard processes on a Lusinian state space with Borel semigroups.)

Let $M \subset]0,\zeta[$ be a homogeneous optional set that is closed in $]0,\zeta[$. It was shown in [5] that there exists a dual object $\hat{M}$ for $\hat{X}$, and [5] was devoted to the study of excursions by X from M or dually by $\hat{X}$ from $\hat{M}$. Let $R = \inf\{t\colon\ t \in M\}$ and $\hat{R} = \inf\{t\colon\ t \in \hat{M}\}$. Then R and $\hat{R}$ are dual exact terminal times for X and $\hat{X}$. Note $R = \infty$ if $R \geq \zeta$ and similarly for $\hat{R}$. The main result of [5] was the existence of measures $P^{x,\ell,y}$ for $x, y \in E$ and $\ell > 0$ which "gave the law of the excursion process starting at x and conditioned to have length ℓ and end at y". Each $P^{x,\ell,y}$ is either a probability or zero. In [5] this was made precise in two situations. Let $G_t = \sup\{s\leq t\colon\ s \in M\}$ and $\check{F}_t = F_{G_t}$. If T is an $(\check{F}_t)$ stopping time and $G = G_T = \sup\{s\leq T\colon\ s \in M\}$, $D = D_T = T + R\circ\theta_T = \inf\{s>T\colon\ s \in M\}$, and $L = D - G$, then the main content of Theorem 7.6 of [5] was that for $F \in bF^*$

$$(3.1)\qquad E^{\mu}[\, F\circ k_R\circ\theta_G \mid F_G,\ T,\ F_{\geq D},\ L,\ X_{D-}\,] = P^{X(G),L,X(D-)}(F)$$

almost surely on $\{G < T,\ L < \infty\}$. Here F_G is the usual σ-field of events before G and $F_{\geq D}$ the σ-field of events after D. See [4] or [5].

In §13 of [5] this was extended to exact terminal times T as follows. Define $G = G_T$, $D = D_T$, and $L = D - G$ as before. Then almost surely on $\{0 < G < T,\ L < \infty\}$

(3.2) $$E^{\mu}[\, F\circ k_R\circ\theta_G \mid \mathcal{F}_G, \mathcal{F}_{\geq D}, L, X_D \,] = P^{X(G),L,X(D-)}[\, F \mid T < \zeta \,]$$

where $P^{x,\ell,y}[\,\cdot \mid T < \zeta\,]$ is just the elementary conditional probability given the event $\{ T < \zeta \}$. Of course, part of the content is that everything makes sense; for example, $P^{X(G),L,X(D-)}[\, T < \zeta \,] > 0$ almost surely on $\{ 0 < G < T,\, L < \infty \}$ is part of (3.2).

The purpose of this section is to study the excursion straddling an arbitrary stopping time T. In light of (2.2), there is no loss of generality in supposing that T is a $(\mathcal{F}^{o*}_{t+})$ stopping time and we let $H \in \mathcal{P}^* \times \mathcal{F}^{o*}$ be such that (2.3) and (2.4) hold. In particular $\omega' \to (H(t,\omega;\omega')-t)^+$ is a $(\mathcal{F}^{o*}_{t+})$ stopping time. As before let $G = G_T$, $D = D_T$, and $L = D - G$. For notational simplicity let $\Lambda = \{ 0 < G < T,\, L < \infty \}$. Define for $G(\omega) < \infty$

(3.3) $$A_\omega(\omega') = (H(G(\omega),\omega;\omega')-G(\omega))^+ .$$

For $\omega \in \Lambda$, A_ω is a $(\mathcal{F}^{o}_{t+})$ stopping time by (2.6), and by (2.3)

$$A_\omega(\theta_G\omega) = T(\omega) - G(\omega)$$

is the age of the excursion straddling T. Finally for $\omega \in \Lambda$ let

(3.4) $$P^{x,\ell,y}[F \mid A_\omega < \zeta] \equiv \int_{\{A_\omega(\omega')<\zeta(\omega')\}} F(\omega')P^{x,\ell,y}(d\omega') \Big/ P^{x,\ell,y}[A_\omega < \zeta],$$

where the ratio is set equal to zero if the denominator vanishes. This is just the elementary conditional probability given $\{\omega' : A_\omega(\omega') < \zeta(\omega') \}$ whenever this event has positive $P^{x,\ell,y}$ probability.

We may now state the main result of this section.

(3.5) THEOREM. *Let* $F \in bF^*$. *Then almost surely on* Λ

$$(3.6)\qquad E^{\mu}[\, F\circ k_R \circ\theta_G \mid F_G,\, F_{\geq D},\, L,\, X_{D-}\,] = P^{X(G),L,X(D-)}[\, F \mid A_\omega < \zeta\,].$$

In particular $P^{X(G),L,X(D-)}(A_\omega<\zeta) > 0$ *almost surely on* Λ, *and if*

$$N = \{(x,\ell,y,\omega):\ P^{x,\ell,y}[\, X_0 = x,\ \zeta = \ell,\ X_{D-} = y \mid A_\omega < \zeta\,] = 1\}$$

then $(X_G(\omega),L(\omega),X_{D-}(\omega),\omega) \in N$ *almost surely on* Λ.

REMARK. The right side of (3.6) is the evaluation of $P^{x,\ell,y}[\, F \mid A_\omega < \zeta\,]$ at $(x,\ell,y) = (X_G(\omega),L(\omega),X_{D-}(\omega))$.

PROOF. The proof of this theorem is very similar to the proofs of (7.6-viii) and (13.7) in [5], and so we will just sketch the argument. Note first that since $((t,\omega),\omega') \to H(t,\omega;\omega')$ is in $P^* \times F^{O*} \subset O^* \times F^{O*}$, $(\omega,\omega') \to A_\omega(\omega')$ is $F_G \times F^{O*}$ measurable. In particular $\omega \to A_\omega(\omega')$ is F_G measurable for each ω'.

Let M_l be the set of strictly positive left endpoints of the intervals contiguous to M and $({}^*P^x,B)$ the Maisonneuve exit system for M. See §5 of [5]. Then $0 < s = G < T < \zeta$ if and only if $s \in M_l$ and $s < T < s + (R\wedge\zeta)\circ\theta_s$ because, M being closed, $T < D$ if $G < T < \zeta$. Therefore one may write for $Z \in bO$, $F \in bF^O$, and all μ

$$E^{\mu}[Z_G\, F\circ\theta_G;\ 0<G<T<\zeta] = E^{\mu} \sum_{s\in M_l} Z_s\, F\circ\theta_s\, 1_{[0,T[}(s)\, 1_{\{T<s+(R\wedge\zeta)\circ\theta_s\}}.$$

But $T(\omega) = H(s,\omega;\theta_s\omega)$, and hence using Maisonneuve's formula (5.4) of [5] and the optionality of $(s,\omega) \to H(\omega,s;\omega')$ this last expression may be written

$$(3.7) \qquad E^{\mu} \int_{]0,T[} Z_s \, {}^{*}P^{X(s)}[\ F;\ H(s,\omega;\cdot) < s + R \wedge \zeta\ ]\ dB_s .$$

Define

$$J(\omega') = J_{s,\omega}(\omega') = (H(s,\omega;\omega')-s)^{+}.$$

Then $J_{s,\omega}$ is an (F^{o}_{t+}) stopping time for each $s > 0$, $\omega \in \Omega$, and $A_{\omega} = J_{G(\omega),\omega}$. Since $R \wedge \zeta \geq 0$, $H(s,\omega;\cdot) < s + R \wedge \zeta$ in (3.7) may be replaced by $J_{s,\omega} < R \wedge \zeta$. A standard monotone class argument shows that $(s,\omega) \to {}^{*}P^{X(s,\omega)}[\ F;\ J_{s,\omega} < R \wedge \zeta\ ]$ is optional, and so defining

$$(3.8) \quad Q^{x,s,\omega}(F) = {}^{*}P^{x}[F | J_{s,\omega} < R\wedge\zeta] = {}^{*}P^{x}[F; J_{s,\omega} < R\wedge\zeta] \big/ {}^{*}P^{x}[J_{s,\omega} < R\wedge\zeta]$$

where the ratio is set equal to zero when the denominator vanishes, we obtain

$$(3.9) \qquad E^{\mu}[\ F\circ\theta_G \mid F_G\ ] = Q^{X(G),G,\omega}(F)$$

almost surely on $\{\ 0 < G < T < \zeta\ \}$. See [3] or [5] for more details of this familiar argument.

Next letting $a \downarrow 0$ in Lemma 7.16 of [5] yields the formula

$$(3.10) \quad {}^{*}P^{x}[F\circ k_R\ \psi(R,X_{R-},X_R)\ ] = {}^{*}P^{x}[P^{x,R,X(R-)}(F)\ \psi(R,X_{R-},X_R)\ ]$$

for $F \in bF^{*}$ and $\psi \in b(B^{+}\times E\times E)$ where B^{+} is the σ-algebra of Borel subsets of $\mathbb{R}^{+}$. Now let $Z \in F_G$, F and $Y \in bF^{o}$, and $\varphi \in b(B^{+}\times E)$. Since $M \subset]0,\zeta[$, $T < \zeta$ and $L = R\circ\theta_G < \zeta\circ\theta_G$ if and only if $L < \infty$ on $\{\ 0 < G < T\ \}$. Hence

$$\Lambda = \{\ 0<G<T,\ L<\infty\ \} = \{\ 0<G<T<\zeta;\ R\circ\theta_G < \zeta\circ\theta_G\ \} = \Lambda_0 \cap \{\ R\circ\theta_G < \zeta\circ\theta_G\ \}$$

where $\Lambda_0 = \{ 0 < G < T < \zeta \}$. Consequently, from (3.9),

$$(3.11)\qquad E^{\mu}[Z\ F\circ k_R\circ\theta_G\ \varphi(L,X_{D-})\ Y\circ\theta_D;\ \Lambda\ ]$$

$$= E^{\mu}\{\ Z\ Q^{X(G),G,\omega}[F\circ k_R\ \varphi(R,X_{R-})\ Y\circ\theta_R;\ R < \zeta];\ \Lambda_0\ \}.$$

Writing $K(x,s,\omega) = {}^*P^x[J_{s,\omega} < R \wedge \zeta\ ]$, since J is a stopping time, one obtains

$$(3.12)\qquad K(x,s,\omega)Q^{x,s,\omega}[F\circ k_R\ \varphi(R,X_{R-})\ Y\circ\theta_R;\ R < \zeta\ ]$$

$$= {}^*P^x[F\circ k_R\ \varphi(R,X_{R-})\ Y\circ\theta_R:\ J < R < \zeta\ ]$$

$$= {}^*P^x[F\circ k_R\ \varphi(R,X_{R-})\ E^{X(R)}(Y);\ J < R\ ].$$

Because J is an (F^{0*}_{t+}) stopping time it is immediate that $\{J < \zeta\}\circ k_R = \{J < R\}$ if $R < \zeta$. See (11.13) of [11], for example. Hence by (3.10) this last expression becomes

$$(3.13)\quad {}^*P^x[P^{x,R,X(R-)}[\ F\ ;J<\zeta]\ \varphi(R,X_{R-})\ \ E^{X(R)}(Y)]$$

$$= {}^*P^x[P^{x,R,X(R-)}[F|J<\zeta]\ P^{x,R,X(R-)}[J<\zeta]\ \varphi(R,X_{R-})\ E^{X(R)}(Y)]$$

$$= {}^*P^x[P^{x,R,X(R-)}[F|J<\zeta]\ \varphi(R,X_{R-})\ Y\circ\theta_R;\ J < R < \zeta\ ]$$

where the last equality uses (3.10) once again. (Note that $J = J_{s,\omega}$ but that ω is fixed in the integration with respect to ${}^*P^x$.) Substituting (3.12) and (3.13) into (3.11) and using (3.9) one more time yields

(3.14) $$E^{\mu}[Z\ F\circ k_R\circ\theta_G\ \varphi(L,X_{D-})\ Y\circ\theta_D;\ \Lambda\ ]$$

$$= E^{\mu}[\ P^{X(G),L,X(D-)}(F|A_{\omega}<\zeta)\ \varphi(L,X_{D-})\ Y\circ\theta_D;\ \Lambda\ ],$$

and this establishes (3.6).

Let h be the indicator of $\{(x,\ell,y,\omega):\ P^{x,\ell,y}(A_{\omega}<\zeta) = 0\ \}$. Then h is $E \times B^{+} \times E \times F_G$ measurable. Recall that $\Lambda_0 = \{0 < G < T < \zeta\}$ and that $\Lambda = \Lambda_0 \cap \{\ R\circ\theta_G < \zeta\circ\theta_G\ \}$. It then follows from (3.9) that

$$E^{\mu}[h(X_G,L,X_{D-},\cdot);\Lambda] = E^{\mu}[h(X_G,R\circ\theta_G,X_{R-}\circ\theta_G,\cdot);\ \Lambda\ ]$$

$$= E^{\mu}\{\int h(X_G,R(w),X_{R-}(w),\cdot)1_{\{R(w)<\zeta(w)\}}Q^{X(G),G,\cdot}(dw);\ \Lambda_0\ \}.$$

But from (3.12) and (3.13),

$$Q^{x,s,\omega}[h(x,R,X_{R-},\omega);\ R < \zeta\ ]$$

$$= K(x,s,\omega)^{-1}\ {}^{*}P^{x}[P^{x,R,X(R-)}(J_{s,\omega}<\zeta)\ h(x,R,X_{R-},\omega);\ R < \zeta\ ]$$

and this last expression is zero when $(x,s) = (X_G,G)$ in view of the definition of h because $J_{G,\omega} = A_{\omega}$. Consequently, almost surely on Λ, $P^{X(G),L,X(D-)}[\ A_{\omega} < \zeta\ ] > 0$.

For the remaining assertion of (3.5) first note that $N \in E \times B^{+} \times E \times F_G$. Let $f(x,\ell,y,\omega)$ be bounded and measurable relative to this product σ-algebra. Since $(X_0,\zeta,X_{\zeta-})\circ k_R\circ\theta_G = (X_G,L,X_{D-})$, it follows from (3.6) that

$$P^{X(G),L,X(D-)}[f(X_0,\zeta,X_{\zeta-},\omega) \mid A_\omega < \zeta\,]$$

$$= E^\mu[f(X_G,L,X_{D-},\cdot) \mid F_G,\ F_{\geq D},\ L,\ X_{D-}\,] = f(X_G,L,X_{D-},\omega)$$

almost surely on Λ. The final assertion of (3.5) follows readily from this.

REMARK. Just as in §8 and §13 of [5] there is a predictable version of Theorem 3.5. We refer the interested reader to [5].

4. Reversing the Excursions

It is assumed throughout this section that X, $\hat{X}$ have dual densities as described in §3. We assume given a point $b \in E$ satisfying

(4.1) b *is regular for itself relative to* X;

(4.2) *all excursions of* X *from* b *begin and end at* b.

The condition (4.2) was discussed in some detail in [5,§10]. In particular, (4.2) was shown to be equivalent to the condition that, up to evanescence,

(4.3) $\{(t,\omega):\ 0<t<\zeta(\omega),\ X_{t-}(\omega)=b\,\} = \{(t,\omega):\ 0<t<\zeta(\omega),\ X_t(\omega)=b\,\}$.

Moreover, the conditions (4.1), (4.2), and (4.3) may be shown to be equivalent to the dual conditions expressed relative to $\hat{X}$.

We shall suppose for convenience that Ω is the space of all maps of $\mathbb{R}^+$ into $E \cup \{\delta\}$ which are right continuous with left limits, admit δ as a trap, and which leave and hit b continuously.

More precisely, the last condition means that (4.3) is to be an identity. Here, (X_t) is the co-ordinate process on Ω. The point is that the measures P^μ, $\hat{P}^\mu$ may be constructed on Ω becuase of the assumptions (4.1) and (4.2). Instead of introducing a separate space $\hat{\Omega}$ for $\hat{X}$ we distinguish the dual process only by its laws $\hat{P}^\mu$ on Ω. Note that Ω is closed under shifts θ_t and killing operators k_t.

We remark that it is an obvious consequence of the strong Markov property that, under (4.1) and (4.2), b is not a holding point.

We adopt from now on the notation of §3 for excursions of X from b, setting

$$M = \{(t,\omega): \ t > 0, \ X_t(\omega) = b \}.$$

Because of (4.3), M is closed in $]0,\zeta[$. Define now Φ: $\Omega \to \Omega$ by

$$(4.4) \qquad (\Phi\omega)(t) = \begin{cases} X_{(G_t(\omega)+D_t(\omega)-t)-}(\omega) & \text{if } 0 < G_t(\omega) < D_t(\omega) < \infty \\ X_t(\omega) & \text{otherwise.} \end{cases}$$

One needs the condition (4.3) in order to check that Φ maps Ω into Ω. The effect of Φ is to reverse (and make right continuous) every excursion of finite length away from b. It was shown in [5,§11] that if X is self dual then

$$\Phi P^\mu = P^\mu \quad \textit{for every} \ \mu.$$

We are going to examine the corresponding result in case X is not necessarily self dual. To begin with, it cannot be expected that $\Phi P^\mu = \hat{P}^\mu$, for the behavior of X until the hitting time of b may be entirely different from that of $\hat{X}$. In general, one may obtain results only in the case $\mu = \varepsilon_b$ so that X and $\hat{X}$ both start at b. Even in this case, there are difficulties in the general case because X and $\hat{X}$ may behave in a very different manner after the last exit time from b. This case is ruled out under the following hypothesis of recurrence on b:

(4.5) $$P^b[\ \sup\{t:\ X_t=b\} < \infty\ ] = 0;$$

$$\hat{P}^b[\ \sup\{t:\ X_t=b\} < \infty\ ] = 0.$$

In §3 and in [5], R was used to denote the debut of M which is $T_b \equiv \inf\{t > 0:\ X_t = b\}$ in the present case. However, in what follows it is convenient to define

(4.6) $$R \equiv \inf\{t > 0:\ X_{t-} = b\}\ .$$

From (4.1) and its dual, $R = T_b$ almost surely P^b and $\hat{P}^b$. In fact, because of (4.3), $R \neq T_b$ if and only if $R = \zeta < \infty = T_b$, and hence $k_R = k_{T_b}$ as operators on Ω. Finally (4.5) implies that $R = T_b$ almost surely $^*P^b$ and $^*\hat{P}^b$. To see this we shall use Maisonneuve's formula [5, Theorem 5.3]. First, we may choose the exit system for M so that the additive functional, ℓ, involved is the local time at b normalized by

(4.7) $$E^x \int_0^\infty e^{-t}\, d\ell_t = E^x(e^{-T_b})\ ,$$

with a similar choice for the dual objects relative to $\hat{M}$. Then

$$E^b \sum_{s\in M_l} 1_{\{R\circ\theta_s \neq T_b\circ\theta_s\}} e^{-s} = {}^*P^b(T_b\neq R)\, E^b \int_0^\infty e^{-t}\, d\ell_t = {}^*P^b(T_b\neq R),$$

where, as usual, M_l is the set of strictly positive left endpoints of the intervals contiguous to M. If $s \in M_l$ and $R\circ\theta_s \neq T_b\circ\theta_s$, then $s < \zeta$, $X_s = b$, and $T_b\circ\theta_s = \infty$. Hence $\sup\{t:\ X_t=b\} = s < \infty$ and so, under (4.5), ${}^*P^b(T_b\neq R) = 0$. Similarly ${}^*\hat{P}^b(T_b\neq R) = 0$. Note that a similar argument shows that

$${}^*P^b(R = \infty) = 0 = {}^*\hat{P}^b(R = \infty).$$

In the remainder of this paper R is defined by (4.6).

(4.8) THEOREM. *Under the hypotheses* (4.1), (4.2), *and* (4.5) *one obtains* $\Phi P^b = \hat{P}^b$.

For the proof of (4.8) we shall use the excursion theory of Ito [6]. As above ℓ_t and $\hat{\ell}_t$ denote local times for X and $\hat{X}$ at b normalized by (4.7) and its dual. This is the same normalization as that used in [6]. Let τ_t, $\hat{\tau}_t$ denote the respective *left* continuous inverses of ℓ_t, $\hat{\ell}_t$. Because of the hypotheses (4.5), $P^b\{\tau_t<\infty\} = 1$ for all $t \geq 0$, and $\hat{P}^b\{\hat{\tau}_t<\infty\} = 1$ for all $t \geq 0$. Under P^b, (τ_{t+}) is a subordinator having a decomposition of the form (P^b a.s. for all $t \geq 0$)

$$\tau_{t+} = mt + \sum_{0<s\leq t} (\tau_{s+}-\tau_s) \tag{4.9}$$

for some constant $m \geq 0$.

Following Ito, define the excursion process (Y_t) for X from b as follows: (Y_t) is the point process with domain

$D_Y = \{(t,\omega): \tau_t(\omega) < \tau_{t+}(\omega)\}$ and values in Ω given by

(4.10) $$Y_t(\omega) = (k_R \circ \theta_{\tau_t})(\omega) \quad \text{if} \quad (t,\omega) \in D_Y.$$

Note that the $\tau_t(\omega)$ with $\tau_t(\omega) < \tau_{t+}(\omega)$ are, up to evanescence, precisely the points in $M_l(\omega)$, and for such a (t,ω), $Y_t(\omega)$ is the portion of the path until the next return to b at $\tau_{t+}(\omega)$. Observe that $\zeta(Y_t) = \tau_{t+}-\tau_t$ and that because of (4.6) $R(Y_t) = \zeta(Y_t)$ if $t \in D_Y$.

It was shown in [6] that (Y_t, P^b) is a Poisson point process, and as such its distribution is governed by its *characteristic measure* n on (Ω, F^o) given by

(4.11) $$n(\Gamma) = E^b \sum_{0<s\leq 1} 1_\Gamma(Y_s)\, 1_{D_Y}(s); \qquad \Gamma \in F^o .$$

In addition, because of the normalization (4.7), n uniquely determines the constant m in (4.9) by the formula [6, Theorem 6.2]

(4.12) $$m = 1 - \int_\Omega [\ 1-\exp(-R)\ ]\, dn.$$

The characteristic measure n is easily identified in terms of Maisonneuve's excursion measure $*P^b$. As we observed above, M_l is P^b-a.s. equal to $\{\tau_s: \ \tau_s < \tau_{s+},\ s > 0\}$. Therefore, given $\Gamma \in F^o$, using (4.11)

$$n(\Gamma) = E^b \sum_{0<s\leq 1} 1_\Gamma(k_R \circ \theta_{\tau_s})\, 1_{\{\tau_s<\tau_{s+}\}} = E^b \sum_{0<u\leq\tau_1} 1_\Gamma(k_R \circ \theta_u)\, 1_{M_l}(u)$$

which, using Maisonneuve's formula, may be written as

$$E^b \int_0^\infty 1_{]0,\tau_1]}(u)\ {}^*P^b(1_\Gamma \circ k_R)\,d\ell_u = {}^*P^b(1_\Gamma \circ k_R)\ E^b\ell(\tau_1) = {}^*P^b(1_\Gamma \circ k_R),$$

since $\ell(\tau_1) = 1$. In other words,

$$\text{(4.13)} \qquad n = k_R({}^*P^b) \quad \text{on} \quad (\Omega, F^o).$$

Let G denote the σ-algebra determined by the excursion process Y. More precisely, G is generated by the counting processes

$$t \to \sum_{0<s\le t} 1_\Gamma(Y_s)\, 1_{D_Y}(s)$$

as Γ varies in F^o. Let G^b be the P^b-completion of G. In the corresponding way, let $\hat{Y}$, $\hat{n}$ be defined relative to the process $\hat{X}$. (It will turn out later that $\hat{Y}$ may be identified with Y.)

In what follows, we let ρ denote the operator of reversal at time R. That is,

$$\text{(4.14)} \qquad (\rho\omega)(t) = \begin{cases} \omega[(R(\omega)-t)-] & \text{if } 0 \le t < R(\omega) < \infty \\ \delta & \text{otherwise.} \end{cases}$$

Observe that if $R(\omega) < \infty$, then by (4.3), $X_{R-}(\omega) = b$ and so $(\rho\omega)(0) = b$ if $0 < R(\omega) < \infty$. Moreover, as noted earlier, $R(Y_t) = \zeta(Y_t)$ for all $t \in D_Y$, so

$$0 = E^b \sum 1_{\{R(Y_t)\ne\zeta(Y_t)\}} 1_{D_Y}(t)$$

which implies

$$\text{(4.15)} \qquad n\{R \ne \zeta\} = 0.$$

The following is the central calculation.

(4.16) LEMMA. $\rho n = \hat{n}$ *on* (Ω, F^o).

PROOF. In the present case letting $a \downarrow 0$ in [5, 6.6] and using [5, 5.19] yields

$$(4.17) \qquad {}^*P^b(R \in d\lambda) = {}^*\hat{P}^b(R \in d\lambda).$$

Suppose $F \in F^o$ and $F \geq 0$. Using the dual of (4.13), [5, 7.16], and (4.17)

$$\hat{n}(F) = {}^*\hat{P}^b(F\circ k_R) = {}^*\hat{P}^b[\hat{P}^{b,R,b}(F)] = \int \hat{P}^{b,\lambda,b}(F) \quad {}^*P^b(R \in d\lambda).$$

But according to [5, 7.6], $\hat{P}^{b,\lambda,b}(F) = P^{b,\lambda,b}(F\circ\rho_\lambda) = P^{b,\lambda,b}(F\circ\rho_\zeta)$. Consequently

$$\hat{n}(F) = \int P^{b,\lambda,b}(F\circ\rho_\zeta) \quad {}^*P^b(R \in d\lambda)$$

$$= {}^*P^b[P^{b,R,b}(F\circ\rho_\zeta)] = {}^*P^b[F\circ\rho\circ k_R] = \rho n(F),$$

proving (4.16).

(4.18) REMARK. It is only in proving (4.16) that we use the hypothesis of a dual density. The remaining arguments of this section require only the weaker duality of resolvents. It is undoubtedly true that (4.16) is also true under the weaker hypothesis.

(4.19) LEMMA. *It is possible to choose the local times* ℓ_t, $\hat{\ell}_t$ *on* Ω *such that* $\ell_t = \ell_t\circ\Phi = \hat{\ell}_t$ *identically.*

PROOF. Let $A_\varepsilon = \{\omega \in \Omega: \sup_{s>0} d(X_s(\omega),b) > \varepsilon\}$, where d is a metric on E compatible with the topology of E. Define N_t^ε by

$$N_t^\varepsilon(\omega) = \sum_{s\in M_l(\omega), s\le t} 1_{A_\varepsilon}(k_R(\theta_s\omega)),$$

so that N_t^ε differs by at most 2 from the number of downcrossings of $[0,\varepsilon]$ before time t by $d(X_s,b)$. Define $B_t(\omega) = \limsup_{\varepsilon\downarrow 0} N_t^\varepsilon(\omega)/n(A_\varepsilon)$, where $\varepsilon \downarrow 0$ through some fixed sequence. Because $\Phi^{-1}(A_\varepsilon) = A_\varepsilon$ and $N_t^\varepsilon(\omega) = N_t^\varepsilon(\Phi\omega)$, $B_t(\Phi\omega) = B_t(\omega)$ for all $t \ge 0$ and all $\omega \in \Omega$. In addition, Theorem 2 of [9] shows that B serves as a local time for X at b. However, $\hat{n}(A_\varepsilon) = \hat{n}(\Phi^{-1}(A_\varepsilon))$, and since $\hat{n}$ is carried by paths which start at b and end at b without hitting b in between, $\hat{n}(\Phi^{-1}A_\varepsilon) = \hat{n}(\rho^{-1}A_\varepsilon) = n(A_\varepsilon)$, using (4.16). It follows that B_t also serves as a local time at b for $\hat{X}$, using Maisonneuve's theorem once again. This establishes (4.19).

We assume from now on that ℓ and $\hat{\ell}$ are chosen so that $\ell = \ell\circ\Phi = \hat{\ell}$ identically. It follows that $\hat{\tau}_t$ is identical to τ_t on Ω, and consequently $\hat{Y} = Y$ identically.

(4.20) LEMMA. $Y\circ\Phi$ *is identical to* ρY.

PROOF. $Y\circ\Phi$ is a point process with domain $\{(t,\omega): \tau_t(\Phi\omega) < \tau_{t+}(\Phi\omega)\}$, which is identical to D_Y. For $(t,\omega) \in D_Y$,

$$Y_t(\Phi\omega) = k_R(\theta_{\tau_t}(\Phi\omega)) = \rho k_R(\theta_{\tau_t}\omega) = \rho(Y_t(\omega)).$$

PROOF OF THEOREM (4.7). As ρY is the image of the process Y under the mapping $\rho: \Omega \to \Omega$ of its range, ρY under $\hat{P}^b$ is also a

Poisson point process with characteristic measure $\rho\hat{n}$, which, by (4.16), is equal to n. Therefore, taking (4.20) into account, $\Phi\hat{P}^b = P^b$ on the σ-algebra G determined by Y. Once we show that $F^o \subset G^b$ (the P^b-completion of G) the proof will be complete. Observe though that, as we saw following (4.10), $\tau_{t+} - \tau_t = \zeta(Y_t)$ if $\tau_t < \tau_{t+}$, so (4.9) yields (P^b a.s. for all $t \geq 0$)

$$\tau_{t+} = mt + \sum_{0<s\leq t} \zeta(Y_s)\, 1_{D_Y}(s).$$

Therefore τ_{t+} is G^b measurable. We may write

$$X_t(\omega) = \begin{cases} Y_s(\omega)(t-\tau_s(\omega)) & \text{if } \tau_s(\omega) \leq t < \tau_{s+}(\omega) \\ b & \text{if } \tau_s(\omega) = t = \tau_{s+}(\omega) \end{cases}$$

and a simple composition argument then shows that X_t is G^b measurable. This completes the proof of (4.7).

References

1. R.M. BLUMENTHAL and R.K. GETOOR. *Markov Processes and Potential Theory*. Academic Press, New York, 1968.

2. R.K. GETOOR. *Markov Processes: Ray Processes and Right Processes*. Lecture Notes in Mathematics *440*. Springer-Verlag, New York, 1975.

3. R.K. GETOOR. Excursions of a Markov process. *Ann.Prob.* *7* (1979), 244-266.

4. R.K. GETOOR and M.J. SHARPE. Markov properties of a Markov process. *Z. Wahrscheinlichkeitstheorie verw. Gebiete* *55* (1981), 313-330.

5. R.K. GETOOR and M.J. SHARPE. Excursions of dual processes. To appear.

6. K. ITO. Poisson point processes attached to Markov processes. *Proc. 6th Berkeley Symp. Math. Stat. Prob. 3*, 225-240. University of California Press, Berkeley, 1971.

7. B. MAISONNEUVE. Exit systems. *Ann. Prob. 3*, (1975), 399-411.

8. B. MAISONNEUVE. On the structure of certain excursions of a Markov process. *Z. Wahrscheinlichkeitstheorie verw. Gebiete 47* (1979), 61-67.

9. B. MAISONNEUVE. Temps local et dénombrements d'excursions. *Z. Wahrscheinlichkeitstheorie verw. Gebiete 52* (1980), 109-113.

10. J.W. PITMAN. Lévy systems and path decompositions. This volume.

11. M.J. SHARPE. *General Theory of Markov Processes*. Forthcoming book.

R.K. GETOOR
Department of Mathematics
University of California-San Diego
La Jolla, CA 92093

M.J. SHARPE
Department of Mathematics
University of California-San Diego
La Jolla, CA 92093

CHARACTERIZATION OF THE LEVY MEASURES OF INVERSE LOCAL TIMES OF GAP DIFFUSION

by

FRANK B. KNIGHT

1. Introduction

Let $X(t)$ be a persistent nonsingular diffusion on an interval Q containing 0 (in the sense of [4]). If we assume a natural scale, then X is characterized by its speed measure $m(dx)$ on Q, and finite endpoints are reflecting. There exists (P^0-a.s.) the local time

$$(1.1)\qquad \ell(t) = \lim_{h_i \to 0+} \frac{1}{m[-h_1, h_2)} \int_0^t I_{[-h_1,h_2)}(X(s))\, ds$$

which is continuous in $t \geq 0$. The right-continuous inverse $\ell^{(-1)}(t) = \inf\{s:\ \ell(s) > t\}$ is an increasing finite process with homogeneous independent increments, and as such (see for example [4, §6.1]) it is characterized by its Lévy measure $n(dy)$ on $(0,\infty)$ through the equation

$$E \exp[\,-\lambda\ell^{(-1)}(t)\,] = \exp[\,-tm_0\lambda - t\int_0^\infty (1-e^{-\lambda y})\, n(dy)\,],\quad \lambda > 0,$$

where $m_0 = m\{0\}$. An interesting problem [4, p. 217] is to characterize the class of all $n(dy)$ which can appear when Q and $m(dx)$

vary. It is shown there that there is a unique measure μ on $[0,\infty)$ such that $n(dy) = dy \int_{0-}^{\infty} e^{-y\gamma}\mu(d\gamma)$, so the problem reduces to characterizing the class of μ.

As posed here, the problem appears very difficult to solve because of the requirement that $m(dx)$ be strictly positive on subintervals of Q. A natural limiting case is to permit $m(dx)$ to vanish on (at most countably many) subintervals, and the corresponding processes $X(t)$ are well-known in the literature under the name "quasi-diffusions" (for example [10] and [11]). Analytically, their treatment is basically the same as for diffusions, except that the elements of the domain of the generator are extended linearly in the intervals where $m(dx)$ vanishes. Also, the probabilistic treatment in terms of time changes of Brownian motion is the same, but since the time changes may have jumps, the paths also have jumps across the intervals where $m(dx)$ vanishes. Accordingly, we will use the simpler term "gap diffusion" for this more general class of diffusions. The analytical foundations (generators, etc.) for treating these processes (in natural scale) were of course laid down in the 1950's and earlier by M.G. Krein and his followers in the context of vibrating strings. The suitability of this class of processes for studying some aspects of diffusion was only gradually recognized, and they do not appear in [4]. But we find here that they provide the right context for the problem mentioned above.

Our approach is simply as follows. We assume at first that 0 is an endpoint of $Q = [0,c)$, and we relax the requirement that the process be persistent. Then the whole problem comes within the general scope of the inverse spectral theorem of M.G. Krein [5]. After a couple of probabilistic transformations (the second of which may be partly new) the solution becomes relatively straightforward. At the same time, it provides a probabilistic approach to the inverse spectral theory, in which the spectral measure is replaced by the Lévy measure. It turns out that the same class of Lévy measures is obtained with or without the two changes of hypothesis just mentioned, and they evidently give the "right" setting for the problem.

It is necessary to begin with some known information about gap diffusion. This is gleaned from a variety of sources and is mostly a straightforward extension of what is in [4] for diffusion. We will be content to state these facts and merely to indicate where and how a complete proof can be found. We assume that $Q=[0,c)$, $0 < c \leq \infty$, $m(dx)$ a measure on Q that is finite on compact subintervals and positive on neighborhoods of 0 (if this last were not the case we would replace 0 by $\inf(\text{support of } m(dx))$). We let $\rho = \sup(\text{support of } m(dx))$. Note that if $m\{\rho\} > 0$ then $\rho < c$. We assume further that $\rho = c$ if $m[0,c) = \infty$, and we say that ρ is *regular* if and only if $m[0,\rho) + \rho < \infty$ (otherwise, ρ is *singular*). Thus $\rho = c$ if ρ is singular, but we may have $\rho = c$ even if ρ is regular provided that $m\{\rho\} = 0$. Let B^+ denote a standard reflected Brownian motion ($B^+(t) \equiv |B(t)|$) and let B_c^+ denote B^+ killed at c ($B_c^+(t) = \Delta$ for $t \geq \inf\{ s: B^+(s) \geq c \}$, where Δ is adjoined as an isolated point). We continue to write B^+ for B_∞^+. Finally, let

$$s_c^+(t,x) = \frac{1}{2}\frac{d}{dx}\int_0^t I_{[0,x)}(B_c^+(s))\, ds$$

denote the local time of B_c^+ (continuous in (t,x), P^0-a.s., by Trotter's Theorem).

THEOREM 1.1. Let $\tau(t)$ denote the right-continuous inverse of the additive functional

(1.2) $$A(t) = \int_{0-}^{\rho^+} s_c^+(t,x)\ m(dx),$$

with $\tau(t) = \infty$ for $t \geq A(\infty)$. Then the process $X(t) = B_c^+(\tau(t))$ (or Δ for $t \geq A(\infty)$) with the usual generated σ-fields and translation operators $\theta_{\tau(t)}$ defines a Hunt process on closure(support of $m(dx)$) $\cup\ \{\Delta\}$ for the probabilities P^x of B_c^+. It is called the gap diffusion on $[0,c)$ with natural scale and speed measure $m(dx)$.

Discussion. There are at least two different ways to show that X is a Hunt process on closure(supp $m(dx)$)$\cup\ \{\Delta\}$. One way is to first show that it is a right process (this is already shown in [3], for example) and then to establish quasi-left-continuity. Since the starting times of excursions of B_c^+ across gaps in the support of $m(dx)$ are totally inaccessible (by the strong Markov property of B_c^+) and the jumps of X occur only when $\tau(t)$ reaches such an excursion time, the quasi-left continuity of X follows by routine stopping time arguments. Alternatively, one can conclude directly by Blumenthal's Theorem and the fact that $X(t)$ has a strongly Feller semi-group as noted in [10, p. 250]. But this requires checking that the analytical results of [10], when transformed to natural scale, pertain to the process $X(t)$. An easy way to do this is to verify that the resolvent operators $R_\lambda f$ of [10] reduce to those of $X(t)$. For this one may begin with the case of regular ρ and $c < \infty$, where it is clear that $R_0 f$ exists for bounded f. Then, by a reasoning familiar for diffusion, the resolvent of $X(t)$ is simply

$$R_0 f(x) = \int_{0-}^{\rho} G_0(x,y)\ f(y)\ m(dy),$$

where

$$G_0(y,x) = G_0(x,y) = (c - y), \qquad 0 \le x \le y < c,$$

and this agrees with [10] for the same $m(dx)$. The resolvent equation then establishes agreement for all $\lambda > 0$, and choosing $f \ge 0$ one can invoke monotone convergence in ρ and c to pass from the restriction of $m(dx)$ to a subinterval of $[0,\rho)$ to the general case. The process $X(t)$ is called in [11, Definition 3.2] the "quasi-diffusion corresponding to the inextensible measure $m(dx)$," but there $m(dx)$ is extended by placing an infinite point mass at c if $c < \infty$ and ρ is regular.

Let us review some basic analytical facts concerning $X(t)$, taken mainly from [10]. To avoid trivialities we assume $\rho \ne 0$. Then there exists a symmetric continuous transition density $p(t,x,y)$ such that $P^x\{ X(t) \in dy \} = p(t,x,y)\, m(dy)$, and $p(t,x,y)$ is linear in y in the intervals where $m(dy) = 0$, $0 \le y < c$. The transition density $p(t,x,y)$ is determined by

$$G_\lambda(x,y) = \int_0^\infty e^{-\lambda t}\, p(t,x,y)\, dt,$$

where $G_\lambda(x,y)$ may be constructed as follows. For any fixed $a \in [0,\rho)$ there exist positive solutions $g_i(x)$ of

$$g_i(x) = g_i(a) + (x-a) \frac{d^+}{dx^+} g_i(a) - \lambda \int_a^x (y-a)\, g_i(y)\, m(dy),$$

$i = 1$ or 2, such that g_1 is nondecreasing on $(0,c)$ with $\frac{d^+}{dx^+} g_1(0) = m\{0\}\, g_1(0)$, g_2 is nonincreasing with $g_2(c) = 0$ (or ≥ 0 if $c = \infty$: when $\rho < \infty$ this means that $\frac{d^+}{dx^+} g_2(\rho) = 0$), and $W = g_1' g_2 - g_2' g_1$ is a strictly positive constant. Then $G_\lambda(x,y)$ is uniquely defined by

$$G_\lambda(x,y) = G_\lambda(y,x) = W^{-1}\, g_1(x)\, g_2(y); \qquad 0 \le x \le y < c.$$

The case of regular ρ and $c < \infty$ gives the Feller end condition

$$0 = (c-\rho)\,\frac{d^-}{dy^-}\,G_\lambda(x,\rho) + (\,\lambda\, m\{\rho\}\,(c-\rho) + 1\,)\,G_\lambda(x,\rho),$$

which becomes $G_\lambda(x,\rho) = 0$ if $c = \rho$, while the case of regular ρ and $c = \infty$ gives

$$0 = \frac{d^-}{dy^-}\,G_\lambda(x,\rho) + \lambda\, m\{\rho\}\,G_\lambda(x,\rho).$$

The second basic result we need is the inverse spectral theorem of M.G. Krein, which may be stated for our purposes as follows.

THEOREM 1.2. There is a unique measure $\sigma(dw)$ on $[0,\infty)$ with $\int_{0-}^{\infty} \frac{1}{1+w}\,\sigma(dw) < \infty$ such that

(1.3) $$G_\lambda(0,0) = \int_{0-}^{\infty} \frac{1}{\lambda+w}\,\sigma(dw), \qquad \lambda > 0.$$

The correspondence induced by (1.3) between pairs $(m(dx),c)$ in Theorem 1.1 and Borel measure $\sigma(dw) \not\equiv 0$ on $[0,\infty)$ with $\int_{0-}^{\infty} \frac{1}{1+w}\,\sigma(dw) < \infty$ is one-to-one and onto.

Discussion. It is the fact that any such measure $\sigma(dw)$ determines a process X that is crucial to our method. For further discussion of this remarkable result, see for example [5], [10], [11], and the remarks below at the start of Section 2. A complete proof can be found in [1, Section 6.6].

The final basic result which we need concerns the relationship of the local time at 0 of $X(t)$ and $G_\lambda(0,0)$. It is a standard consequence of (1.2) that $X(t)$ has local time $s_c^+(\tau(t),x)$ with respect to $m(dx)$, i.e., P^0-a.s.

$$s_c^+(\tau(t),x) = \frac{d^+}{dm^+} \int_0^t I_{[0,x)}(X(s))\, ds. \tag{1.4}$$

We continue the notation $\ell(t) = s_c^+(\tau(t),0)$ in accordance with (1.1), and $\ell^{(-1)}(t)$ for the right-continuous inverse. If we permit the value $\ell^{(-1)}(t) = \infty$, then $\ell^{(-1)}(t)$ is an increasing process with homogeneous independent increments for P^0, and it follows from the general theory of such processes (see for example J.F.C. Kingman [6]) that there exists a unique triple $(m_0,b,n(dy))$, $0 \le m_0$, $0 \le b$, $n(dy)$ a measure on $(0,\infty)$, such that

$$E^0\exp[-\lambda\ell^{(-1)}(t)] = \exp\{ -t[m_0\lambda + b + \int_0^\infty (1-e^{-\lambda y})\, n(dy)] \}. \tag{1.5}$$

In the present case we have $b = c^{-1}$, i.e. the lifetime of $\ell^{(-1)}(t)$ is exponential with parameter c^{-1}. To see this one need only observe that the lifetime of $\ell^{(-1)}(t)$ is the local time of the first excursion from 0 to reach $[c,\infty)$. The Poisson rate of such excursions, in local time at 0, is the same for all $X(t)$ as it is for B^+, namely c^{-1} (see [12] and, for another interpretation via the passage times to 0 of the processes $s^+(\ell^{(-1)}(t),x)$ in parameter x, see [8, Theorems 4.3.6 and 5.3.23]). We also have $m_0 = m\{0\}$, which is contained below in Theorem 2.1. The connection we need between $\ell(t)$ and $G_\lambda(0,0)$ is

THEOREM 1.3.

$$E^0\exp(-\lambda\ell^{(-1)}(t)) = \exp(-t/G_\lambda(0,0)), \qquad \lambda > 0.$$

Discussion. To prove this, one need only go through the argument of [4, Section 5.4], use the formula given above for $G_\lambda(0,0)$, and make a change of variables to obtain

$$G_\lambda(0,0) = E^0 \int_0^\infty e^{-\lambda t}\, d\ell(t) = E^0 \int_0^\infty e^{-\lambda \ell^{(-1)}(t)}\, dt.$$

This identifies the factor of t on the right side of (1.5) as $G_\lambda^{-1}(0,0)$. In fact, by combining Theorems 1.2 and 1.3 with (1.5) we have immediately the following equation, which provides the starting point for our subsequent analysis:

$$(1.6) \qquad [\, m_0\lambda + c^{-1} + \int_0^\infty (1-e^{-\lambda y})\, n(dy)\,] \int_{0-}^\infty \frac{1}{\lambda+w}\, \sigma(dw) = 1.$$

The reader who is interested only in the characterization problem, and who can accept the first sentence of Section 3, may omit Section 2.

2. Two Transformations

In view of Theorems 1.2 and 1.3, the transforms $E^0\exp[-\lambda\ell^{(-1)}(t)]$ determine $(m(dx),c)$ uniquely. It is to be noted that, in the case when X is persistent, it follows (in a sense) that the zero set of a generic path alone determines $\frac{d}{dm}\frac{d}{dx}$. In the first place, it is clear from the strong law of large numbers that if $(a,b) \subset (c,d)$

$$P^0\{\ \frac{n(a,b)}{n(c,d)} = \lim_{t\to\infty} \frac{\#\{Z^+ :\ Z^+ \subset [0,t),\ a < |Z^+| < b\ \}}{\#\{Z^+ :\ Z^+ \subset [0,t),\ c < |Z^+| < d\ \}}\ \} = 1,$$

where Z^+ denote intervals of $\{\ t : X(t) > 0\ \}$ and $|Z^+|$ their duration (as in [4, §6.3]). Thus $n(dy)$ is P^0-a.s. determined up to a constant factor. However, this factor cannot be determined by the zero set, for if it could then $(m(dx),c)$ would be determined from Theorem 1.2, while in fact $X(t)$ and $k\,X(t)$ have the same zero set but $k\,X(t)$ has pair $(k^{-1}dm(k^{-1}x),kc)$.

On the other hand, once the factor is chosen then $(m(dx),c)$ is completely determined. To see this, suppose first that $n(0,\infty) = \infty$. Then $n(dy)$ together with the zero set determines $\ell(t)$ by formula 2 b) of [4, §6.3], whose proof remains valid (we will see later that $n(dy)$ is absolutely continuous). Then $n(dy)$ is the Lévy measure of $\ell^{(-1)}(t)$ and $m\{0\}$ is determined from $m\{0\} = \int_0^t I_{\{0\}}(X(s))ds/\ell(t)$. It is shown below that $m\{0\} = m_0$. Hence by Theorem 1.2, $(m\{0\},n(dy))$ determines $(m(dx),c)$. If $n(0,\infty) < \infty$, on the other hand, there must be a gap at 0 and $m\{0\} > 0$. Then we can choose $m\{0\} > 0$ arbitrarily, thus determining $\ell(t)$ as before. In turn this determines $n(dy)$ since, by the strong law of large numbers,

$$P^0\{\ n(b,\infty) = \lim_{t\to\infty} \frac{1}{t}\ \#\{Z^+\colon\ Z^+ \subset [0,\ell^{(-1)}(t)),\ b < |Z^+|\}\ \} = 1.$$

Once again, the pair $(m\{0\},n(dy))$ determines $(m(dx),c)$. We can summarize the situation by the observation that the zero set determines $\frac{d}{dm}\frac{d^+}{dx^+}$ P^0-a.s. in a form invariant under scale change. Of course, this determination breaks down for diffusion on $(-\infty,\infty)$ since there the zero set does not discriminate positive and negative excursions.

We turn now to the connection (or rather, its absence) of m_0 and b with $n(dy)$ when the scale dx is fixed in advance. The following theorem shows that in characterizing $n(dy)$ we can assume $m_0 = 1$.

THEOREM 2.1. If $(m_0,b,n(dy))$ is possible in (1.6) (with $b=c^{-1}$ and corresponding speed measure $m(dx)$), then $(m_1,b,n(dy))$ is possible for all $m_1 > 0$ and if $0 \in \text{supp}\,(I_{(0,\infty)}(x)m(dx))$, then $m_1 = 0$ is possible. In any case, $m_0 = m\{0\}$ in the corresponding speed measure.

PROOF. We write $A(t)$ from (1.2) in the form $A(t) = A_0(t) + A_1(t)$, where $A_0(t) = \varepsilon s_c^+(t,0)m\{0\}$, $0 < \varepsilon < 1$, and $m(dx)$ is the measure corresponding to $(m_0,b,n(dy))$. Then if $m_\varepsilon(dx)$ denotes the same measure for the process $X_1(t) = B_c^+(A_1^{(-1)}(t))$, we have $m_\varepsilon\{0\} = (1-\varepsilon)m\{0\}$. Since for each t, $s_c^{+(-1)}(t,0)$ is P^0-a.s. a time of increase of $A(t)$, and $\ell(t) = s_c^+(A^{(-1)}(t),0)$ as before, it follows that

$$\ell^{(-1)}(t) = A(s_c^{+(-1)}(t,0)) = A_1(s_c^{+(-1)}(t,0)) + \varepsilon\, m\{0\}t ,$$

and

$$\ell_1^{(-1)} = A_1(s_c^{+(-1)}(t,0))$$

is the inverse local time at 0 of X_1. Thus clearly X and X_1 have the same $n(dy)$ and b, but if $m_0(\varepsilon)$ denotes m_0 for X_1 then $m_0 = m_0(\varepsilon) + \varepsilon m\{0\}$. Since

$$\lim_{\varepsilon\to 1} \int_0^t I_{\{0\}}(X_1(s))\, ds = 0,$$

and since it is well known ([5, Section 6]) that $m_0(\varepsilon)t$ is the absolutely continuous component of $\ell_1^{(-1)}(t)$, it follows easily that $\lim_{\varepsilon\to 1} m_0(\varepsilon) = 0$. Hence $m_0 = m\{0\}$, and it also follows that for $m_0 > 0$ we can obtain any smaller positive m_0 without changing

$(b,n(dy))$. The same argument applied to the process $X(t)$ determined by a speed measure with point mass k at 0, say $m_k(dx) = k\delta_0 + I_{(0,\infty)}(x)m(dx)$, for any $k > 0$, shows that any positive m_0 may be obtained. But to permit $k = 0$ we must assume that $\ell(t)$ remains well-defined, hence $0 \in \text{supp}\,(I_{(0,\infty)}(x)m(dx))$.

The next problem is to determine the range of possible b for given $n(dy)$. We assume without loss of generality that $m\{0\} = 1$. Our theorem has three parts.

THEOREM 2.2. (i) Suppose that $(1,b,n(dy))$ is possible in (1.6) with $b = c^{-1} > 0$. For $c \le \alpha < \infty$ we define a new pair (m_α,c_α) by

$$\begin{aligned} dm_\alpha(x) &= \left(\frac{\alpha}{\alpha+x}\right)^2 dm\left(\frac{\alpha}{\alpha+x}\,x\right), \qquad 0 \le x \le \frac{\alpha}{\alpha-\rho}\,\rho\ , \\ c_\alpha &= \frac{\alpha}{\alpha-c}\,c \qquad (c_\alpha = \infty \ \text{ if } \ \alpha = c). \end{aligned} \tag{2.1}$$

Then (m_α,c_α) corresponds to $(1,b_\alpha,n(dy))$ with $b_\alpha = (b\alpha-1)\alpha^{-1}$. Thus b_α may be any value in $[0,b)$.

(ii) Suppose only that $(1,b,n(dy))$ is possible in (1.6). Then for $0 < \beta < \infty$ we define

$$\begin{aligned} dm_\beta(x) &= \left(\frac{\beta}{\beta-x}\right)^2 dm\left(\frac{\beta}{\beta-x}\,x\right), \qquad 0 \le x \le \frac{\beta}{\beta+\rho}\,\rho\ , \\ c_\beta &= \frac{\beta}{\beta+c}\,c \qquad (c_\beta = \beta \ \text{ if } \ c = \infty). \end{aligned} \tag{2.2}$$

Then the pair (m_β,c_β) corresponds to $(1,b_\beta,n(dy))$ with $b_\beta = (1+b\beta)\beta^{-1}$. Thus b_β may be any value in (b,∞).

(iii) The Feller boundary type of ρ is preserved under (2.1) and (2.2) if ρ is a natural boundary, but this need not hold for ρ regular, exit-not-entrance, nor entrance-not-exit. On the other hand, the grenz-punkt, grenz-kries classification of H. Weyl is preserved, and along with it Feller's classification into active and semiactive [2, p. 468].

PROOF. The idea for the proof of (i) is to replace B^+ in Theorem 1.1 by B^+ conditioned never to reach $[\alpha,\infty)$. Such a process is readily obtained by the excision procedure of [7]: one simply excises (or deletes) the excursions of B^+ which reach $[\alpha,\infty)$. The result is a Hunt process by [7], and since its paths are continuous it is a diffusion on $[0,\alpha)$. Then the identification is readily completed: Brownian motion on $[0,\infty)$ conditioned not to reach 0 is the 3-dimensional Bessel process $r(t)$ (see [8, Lemma 5.2.8] for example) so this excised process, which we denote B_+^α, is such that $\alpha - B_+^\alpha$ is an $r(t)$ reflected at α. In other words, B_+^α is the diffusion on $[0,\alpha)$ with generator $\frac{1}{2}\left(\frac{d^2}{dx^2} - \frac{2}{\alpha-x}\frac{d}{dx}\right)$ and the reflecting barrier end condition $F'(0) = 0$.

Let

$$s_c^*(t,x) = \frac{1}{2}\frac{d}{dx}\int_0^{t\wedge T_\alpha(c)} I_{[0,x)}(B_+^\alpha(s))\,ds,$$

where $T_\alpha(c)$ is the passage time of B_+^α to $[c,\infty)$, $0 \le x < c$. Then using B_+^α for B^+ in Theorem 1.1 simply means replacing s_c^+ by s_c^* in (1.2). We define a corresponding process $X^*(t)$ by

$$X^*(t) = B_+^\alpha(\tau^*(t)), \qquad 0 \le t < \zeta^* = \int_{0-}^{\rho^+} s_c^*(T_\alpha(c),x)\,m(dx) \quad ,$$

where $\tau^*(t)$ is the inverse function of

$$(2.3)\qquad A^*(t) = \int_{0-}^{\rho^+} s_c^*(t,x)\, m(dx).$$

Then since none of the deleted excursions contributed to $\ell^{(-1)}(t)$ it is readily seen that the local time of X^* at 0 with respect to $m(dx)$, namely $\ell^*(t) = s_c^*(\tau^*(t),0)$, determines the same Lévy measure $n(dy)$ as $\ell(t)$. In fact, if we consider the excursions from 0 of B^+ parametrized by $s^+(t,0)$ it is well-known that they comprise a Poisson point process and the mean rate of excursions with maximum in (x_1,x_2), $0 < x_1 < x_2 < c$, is $x_1^{-1} - x_2^{-1}$ (as above following (1.5)). Since the time changes do not affect the excursion maxima, the same is then true of both $X(t)$ and $X^*(t)$, using their respective local times at 0 as parameter, except that now the Poisson processes are killed at the first excursion to $[c,\infty)$. However, the local time of this killing time is clearly smaller for X than for X^* if the first excursion of B^+ reaching $[c,\infty)$ also reaches $[\alpha,\infty)$. The rate of excursions of X^* reaching $[c,\infty)$ is only $c^{-1} - \alpha^{-1}$, whence for X^* the parameter b is replaced by

$$(2.4)\qquad b_\alpha = (c^{-1} - \alpha^{-1}) = (b\alpha - 1)\alpha^{-1} .$$

The problem remains, since dx is not a natural scale for X^*, to express X^* as a time and scale change of a new B^+-process without changing b_α or $n(dy)$. This is a standard diffusion problem. We first express B_+^α in terms of the new B^+ in the scale s and speed measure ν given by

$$(2.5)\qquad s(x) = \frac{\alpha x}{\alpha - x}, \quad \frac{d\nu}{dx} = 2\left(\frac{\alpha - x}{\alpha}\right)^2, \qquad 0 \le x < \alpha.$$

It follows that the inverse of $s(x)$ is $u(s) = s\alpha(\alpha + s)^{-1}$, and further we obtain

$$\frac{d}{d\nu}\frac{d}{ds} = \frac{d}{d\nu}\left(\left(\frac{\alpha-x}{\alpha}\right)^2 \frac{d}{dx}\right)$$

$$= \frac{1}{2}\left(\frac{\alpha}{\alpha-x}\right)^2 \frac{d}{dx}\left(\left(\frac{\alpha-x}{\alpha}\right)^2 \frac{d}{dx}\right) = \frac{1}{2}\left(\frac{d^2}{dx^2} - \frac{2}{\alpha-x}\frac{d}{dx}\right)$$

which is the required generator of B_+^α. It follows that B_+^α is equivalent to $u(B^+(\tau_\alpha(t)))$, where $\tau_\alpha(t)$ is the inverse of

$$A_\alpha(t) = \int_0^t \frac{1}{2}\frac{d\nu}{ds}(B^+(r))dr = \int_0^t \alpha^4(1 + B^+(r))^{-4}dr.$$

We next express the local time $s^*(t,x)$ of B_+^α with respect to $2dx$ in terms of this B^+.

$$(2.6) \qquad s^*(t,x) \equiv \frac{1}{2}\frac{d}{dx}\int_0^t I_{[0,x)}(u(B^+(\tau_\alpha(s))))\, ds$$

$$= \frac{1}{2}\frac{d}{dx}\int_0^{\tau_\alpha(t)} I_{[0,s(x))}(B^+(r))\, dA_\alpha(r)$$

$$= \frac{d}{dx}\int_0^{s(x)} \alpha^4(1 + y)^{-4}\, s^+(\tau_\alpha(t),y)\, dy$$

$$= \alpha^{-2}(\alpha - x)^2\, s^+(\tau_\alpha(t),s(x)),$$

where s^+ is the local time of B^+.

Therefore, the A^* of (2.3) becomes

$$(2.7) \qquad A^*(t) = \alpha^{-2}\int_{0-}^{\rho^+} (\alpha - x)^2\, s^+_{s(c)}(\tau_\alpha(t),s(x))\, m(dx),$$

where $s^+_{s(c)}$ is the local time of $B^+_{s(c)}$ (i.e. of B^+ killed at $s(c)$). It follows from this that $X^*(t)$ becomes

$$X^*(t) = u(B^+((A^*A_\alpha)^{(-1)}(t))),$$

where

$$A^*A_\alpha(t) = \alpha^{-2} \int_{0-}^{\rho^+} (\alpha - x)^2 \, s^+_{s(c)}(t,s(x)) \, m(dx)$$

$$= \int_{0-}^{\infty} \alpha^2(\alpha + y)^{-2} \, s^+_{s(c)}(t,y) \, dm(u(y)).$$

Therefore the speed measure of X^* is

$$(2.8) \qquad (\frac{\alpha}{\alpha + x})^2 \, dm(u(x)).$$

We observe finally that $\ell^*(t)$, the local time at 0 of X^* with respect to $m(dx)$, is the same as that of $B^+_{s(c)}((A^*A_\alpha)^{(-1)}(t))$ with respect to $dm(u(x))$, which is also the same at 0 as the local time with respect to the speed measure (2.8) of $B^+_{s(c)}((A^*A_\alpha)^{(-1)}(t))$. Thus we have expressed ℓ^* in the form of a local time at 0 of a diffusion in natural scale. By (2.4), (2.5), and (2.8), this corresponds to the pair (m_α, c_α) of (2.1).

We turn next to Theorem 2.2(ii). To show this we simply invert the transformation (2.1) by setting, for $0 < \beta < \infty$,

$$dm(x) = (\frac{\beta}{\beta + x})^2 \, dm_\beta(\frac{\beta x}{\beta + x})$$

where (dm,c) is determined by $(1,b,n(dy))$. It follows that

$$(2.9) \qquad dm_\beta(y) = (\frac{\beta}{\beta - y})^2 \, dm(\frac{\beta y}{\beta - y}), \qquad 0 \le y < \frac{\beta\rho}{\beta + \rho} .$$

Thus dm_β is the speed measure of a transient diffusion, and if we consider this process killed at $c_\beta = \beta c(\beta + c)^{-1}$ then we can apply part (i) with $\alpha = \beta$ to recover (dm,c). It follows that $n(dy)$ is preserved, and the new value of b is $b_\beta = c_\beta^{-1} = (1 + b\beta)\beta^{-1}$ as asserted.

Turning finally to (iii), we note that if $(m(dx),c)$ is given, $c < \infty$, and $m(dx)$ is strictly positive on $(0,\rho)$, then with α as in (i) we can write, for F in the appropriate domain,

$$\frac{d}{dm}\frac{d^+}{dx^+} F = \frac{\alpha}{\alpha-x}\frac{d}{dm}\left(\frac{\alpha-x}{\alpha}\frac{d^+}{dx^+}\left(\frac{\alpha}{\alpha-x} F\right)\right), \qquad 0 \le x < c.$$

This is entirely analogous to the identity $D_m D_x F = x^{-1} D_m (x^2 D_x (x^{-1} F))$ which is (4.8) of [2], so we leave to the reader the verification that the left side is continuous if and only if the right side is. This expresses the operator $\frac{d}{dm}\frac{d^+}{dx^+}$ in the form A_ψ of Feller [2, (1.8)] with $\psi = \alpha^{-1}(\alpha - x)$. It is easy to check that the measure m_α of (i) is the speed measure of the related operator A^* of [2, (4.3)], which makes the assertions relative to (2.1) a special case of [2, Theorem 4.2]. It follows immediately (if $m(dx) > 0$) that the inverse transformation (ii) also has the asserted properties. Extensibility of Feller's results to permit $m(dx) = 0$ is probably a foregone conclusion, but it is easier to check the assertions of (iii) directly in this case, which will be omitted.

3. Characterizations of the Lévy measures

From now on, in characterizing the class of Lévy measures possible in (1.6), we assume without loss of generality that $m_0 = 1$ and $c^{-1} = 0$. By Theorem 1.2 the class of $n(dy)$ is then obtained from (1.6) by allowing $\sigma(dw)$ to be any measure on $[0,\infty)$ with $\int_{0-}^{\infty} \frac{1}{1+w}\sigma(dw) < \infty$, compatible with these assumptions. These may be expressed in terms of $\sigma(dw)$ as follows. Since $1 - e^{-\lambda y}$ is monotone, it follows from the finiteness of $\int_0^\infty (1 - e^{-\lambda y})n(dy)$ and an integration by parts that

$$\int_0^\infty (1 - e^{-\lambda y})\, n(dy) = \lambda \int_0^\infty e^{-\lambda y}\, n[y,\infty)dy.$$

Thus (1.6) becomes

$$\text{(3.1)} \qquad [\, m_0 + (c\lambda)^{-1} + \int_0^\infty e^{-\lambda y}\, n[y,\infty)\, dy\,][\, \int_{0-}^\infty \frac{\lambda}{\lambda+w}\, \sigma(dw)] = 1.$$

Letting $\lambda \to \infty$ we obtain

$$\text{(3.2)} \qquad m_0 = \sigma[0,\infty)^{-1} \qquad (\text{or } 0 \text{ if } \sigma[0,\infty) = \infty).$$

On the other hand, letting $\lambda \to 0$ in (1.6), it follows by dominated convergence that

$$\text{(3.3)} \qquad c^{-1} = [\, \int_{0-}^\infty \frac{1}{w}\, \sigma(dw)\,]^{-1} \qquad (\text{or } 0 \text{ if } \sigma\{0\} > 0).$$

Thus our two assumptions become

$$\text{(3.4)} \qquad \sigma[0,\infty) = 1, \qquad \lim_{\lambda\to 0} \int_{0-}^\infty \frac{1}{\lambda+w}\, \sigma(dw) = \infty .$$

Finally, although it will not be used below, we obtain in this case from (3.1)

$$\text{(3.5)} \qquad \lambda \int_0^\infty e^{-\lambda y}\, n[y,\infty)\, dy = \lambda\, [\, \int_{0-}^\infty \frac{\lambda}{\lambda+w}\, \sigma(dw)\,]^{-1} - \lambda$$

$$= \lambda\, [\, \int_{0-}^\infty \frac{w}{\lambda+w}\, \sigma(dw)\,]\, [\, 1 - \int_{0-}^\infty \frac{w}{\lambda+w}\, \sigma(dw)\,]^{-1} .$$

Letting $\lambda \to \infty$, it follows by monotone convergence that

$$n(0,\infty) = \int_0^\infty w\,\sigma(dw). \tag{3.6}$$

Now since

$$\int_{0-}^\infty \frac{1}{\lambda+w}\,\sigma(dw) = \int_0^\infty e^{-\lambda y}\left[\int_{0-}^\infty e^{-yw}\,\sigma(dw)\right] dy,$$

we can write (3.1), under assumptions (3.4), as

$$\lambda^{-1} = \left[1 + \int_0^\infty e^{-\lambda y}\, n[y,\infty)\, dy\right] \int_0^\infty e^{-\lambda y}\left[\int_0^\infty e^{-yw}\,\sigma(dw)\right] dy. \tag{3.7}$$

The left side is the Laplace transform of 1, whence by inversion we obtain

$$1 = \int_0^\infty e^{-yw}\,\sigma(dw) + \int_0^y \left[\int_{0-}^\infty e^{-(y-z)w}\,\sigma(dw)\right] n[z,\infty)\, dz. \tag{3.8}$$

The first term on the right of (3.8) is differentiable, and it follows readily that the second term has derivative

$$n[y,\infty) - \int_0^y \int_{0-}^\infty w e^{-(y-z)w}\,\sigma(dw)\, n[z,\infty)\, dz.$$

Thus we obtain

$$n[y,\infty) = \int_0^\infty w e^{-yw}\,\sigma(dw) + \left[\int_0^\infty w e^{-yw}\,\sigma(dw)\right] * n[y,\infty) \tag{3.9}$$

where $*$ denotes convolution. This is a standard renewal equation.

We now set

$$F(\lambda) = \int_0^\infty \frac{w}{\lambda + w}\,\sigma(dw) < 1$$

and

$$G(\lambda) = \int_0^\infty e^{-\lambda y}\, n[y,\infty)dy.$$

Again taking transforms in (3.9) we obtain

$$G(\lambda) = F(\lambda)(1 - F(\lambda))^{-1}\ , \tag{3.10}$$

and consequently

$$G(\lambda) = \sum_{n=1}^{\infty} (F(\lambda))^n\ .$$

It follows that

$$n[y,\infty) = \sum_{n=1}^{\infty} f^{n*}(y),$$

where

$$f(y) = \int_0^\infty we^{-yw}\,\sigma(dw).$$

We are now ready to state and prove

THEOREM 3.1. The class $\{n(dy)\}$ of Lévy measures of persistent gap diffusion on $[0,\infty)$, reflected at 0 as in Theorem 1.1, consists of all

$$n(dy) = [\int_0^\infty e^{-yz}\,\mu(dz)\,]\ dy$$

with measures $\mu(dz) \geq 0$ on $(0,\infty)$ such that $\int_0^\infty \frac{1}{x(1+x)}\,\mu(dx) < \infty$.

REMARK. It is easily verified that the last condition is simply the necessary and sufficient condition that $n(dy)$ of the above form be a Lévy measure. Thus, in a sense, our characterization problem has the trivial solution.

PROOF. As seen above, the persistency requirement $b = 0$ does not restrict the class $\{n(dy)\}$. The class of possible $F(\lambda)$ in (3.10) consists of all $F(\lambda)$ with $\sigma[0,\infty) = 1$ and $\int_{0-}^{\infty} \frac{1}{w}\sigma(dw) = \infty$. According to what we wish to prove, we should have

$$G(\lambda) = \int_0^\infty e^{-\lambda y}\left[\int_0^\infty e^{-yz}\frac{\mu(dz)}{z}\right]dy = \int_0^\infty \frac{1}{\lambda+z}\frac{\mu(dz)}{z} ,$$

so that $G(\lambda)$ is the Stieltjes transform of $z^{-1}\mu(dz)$. We need a characterization of Stieltjes transforms, which is in [9, Appendix A4], as follows.

LEMMA 3.1. A function of complex λ, continuous for $\mathrm{Re}\,\lambda > 0$, has the form

$$H(\lambda) = \int_0^\infty \frac{1}{\lambda+x}\,\nu(dx) + \gamma, \quad \nu(dx) \geq 0, \quad \gamma \geq 0,$$

if and only if

i) $H(\lambda)$ is holomorphic in $\{\mathrm{Im}\,\lambda < 0\}$,

ii) $\mathrm{Im}\,H(\lambda) \geq 0$ in $\{\mathrm{Im}\,\lambda < 0\}$, and

iii) $H(\lambda)$ is both holomorphic and non-negative real on the axis $0 < \lambda < \infty$.

Using this result and (3.10) we first show that $G(\lambda)$ satisfies these conditions along with $F(\lambda)$. Indeed, since

$$\operatorname{Im} F(\lambda) = -\operatorname{Im} \lambda \int_0^\infty |\lambda + w|^{-2} w\sigma(dw),$$

if $\operatorname{Im} \lambda < 0$ then either $\operatorname{Im} F(\lambda) \neq 0$ or $\sigma(dw) \equiv 0$ on $(0,\infty)$ (i.e. $\sigma\{0\} = 1$). In either case $(1 - F(\lambda))^{-1}$ is holomorphic in $\{\operatorname{Im} \lambda < 0\}$, and along with it $G(\lambda)$.

Next we write

$$\operatorname{Im} G(\lambda) = |1 - F(\lambda)|^{-2} \operatorname{Im}(F(\lambda)((1 - \operatorname{Re} F(\lambda)) + i \operatorname{Im} F(\lambda)))$$

$$= |1 - F(\lambda)|^{-2} \operatorname{Im} F(\lambda)$$

which is ≥ 0 along with $\operatorname{Im} F(\lambda)$. Also, since $0 \leq F(\lambda) < 1$ for $\lambda > 0$, (iii) follows immediately for $G(\lambda)$. Thus by Lemma 3.1 we have

$$G(\lambda) = \int_{0-}^\infty \frac{1}{\lambda+x} \nu(dx) + \gamma .$$

But since

$$\lim_{\lambda\to\infty} F(\lambda) = 0 = \lim_{\lambda\to\infty} G(\lambda) \quad ,$$

we have $\gamma = 0$. Writing

$$G(\lambda) = \int_0^\infty e^{-\lambda y} \left[\int_{0-}^\infty e^{-yx} \nu(dx) \right] dy,$$

we obtain

$$n[y,\infty) = \int_{0-}^\infty e^{-yx} \nu(dx) \quad ,$$

and thus

$$\nu\{0\} = \lim_{y\to\infty} n[y,\infty) = 0.$$

Then, setting $\mu(dz) = z\nu(dz)$, we obtain

$$n(dy) = [\int_0^\infty e^{-yz}\,\mu(dz)\,]\,dy.$$

Next, we must impose the condition

$$(3.11)\quad \infty = \lim_{\lambda\to 0}\int_{0-}^\infty \frac{1}{\lambda+w}\,\sigma(dw)$$

$$= \lim_{\lambda\to 0}\frac{1}{\lambda}\,[\int_{0-}^\infty \sigma(dw) - \int_{0-}^\infty \frac{w}{\lambda+w}\,\sigma(dw)\,]$$

$$= \lim_{\lambda\to 0}\frac{1}{\lambda}\,(1-F(\lambda)) = \lim_{\lambda\to 0}\frac{1}{\lambda}[\,1-\frac{G(\lambda)}{1+G(\lambda)}\,] = \lim_{\lambda\to 0}\frac{1}{\lambda[1+G(\lambda)]}\;.$$

Therefore, we have

$$(3.12)\qquad 0 = \lim_{\lambda\to 0}\lambda(1+G(\lambda))$$

$$= \lim_{\lambda\to 0}\lambda\int_0^\infty e^{-\lambda y}\,n[y,\infty)\,dy$$

$$= \lim_{\lambda\to 0}\lambda\int_0^\infty e^{-\lambda y}\,[\int_0^\infty e^{-yz}\,z^{-1}\,\mu(dz)\,]\,dy$$

$$= \lim_{\lambda\to 0}\lambda\int_0^\infty \frac{1}{z(\lambda+z)}\,\mu(dz).$$

By dominated convergence this is equivalent to finiteness of $\int_0^\infty (x(1+x))^{-1}\,\mu(dx)$.

Suppose, conversely, that we begin with

$$G(\lambda) = \int_0^\infty \frac{1}{z(\lambda+z)}\, \mu(dz),$$

with $\mu(dz)$ satisfying this condition, and set $F(\lambda) = G(\lambda)(1+G(\lambda))^{-1}$ as in (3.10). We wish to check that

$$F(\lambda) = \int_0^\infty \frac{w}{\lambda+w}\, \sigma(dw)$$

with $\sigma(dw)$ as specified at the start of the proof. Now

$$\text{Im } G(\lambda) = -(\text{Im } \lambda) \int_0^\infty z^{-1}|\lambda + z|^{-2}\, \mu(dz)$$

which is non-zero for $\text{Im } \lambda \neq 0$ unless $\mu \equiv 0$, so in either case $F(\lambda)$ is holomorphic in $\{\text{Im } \lambda < 0\}$ along with $G(\lambda)$. Also,

$$\begin{aligned}\text{Im } F(\lambda) &= |1 + G(\lambda)|^{-2}\, \text{Im}(G(\lambda)((1 + \text{Re } G(\lambda)) - i \text{ Im } G(\lambda)) \\ &= |1 + G(\lambda)|^{-2}\, \text{Im } G(\lambda),\end{aligned}$$

which is ≥ 0 along with $\text{Im } G(\lambda)$. And since $F(\lambda)$ is obviously ≥ 0 and holomorphic in $0 < \lambda < \infty$, we can write $F(\lambda)$ as a Stieltjes transform. But $\lim_{\lambda\uparrow\infty} F(\lambda) = 0$ is clear, hence we have

$$F(\lambda) = \int_0^\infty \frac{w}{\lambda+w}\, \sigma(dw)$$

for a unique $\sigma(dw)$ on $(0,\infty)$. Since obviously $F(\lambda) < 1$, we have

$$\sigma(0,\infty) = \lim_{\lambda\to 0} F(\lambda) \leq 1,$$

and so we can define $\sigma\{0\} = 1 - \sigma(0,\infty)$ to obtain $\sigma[0,\infty) = 1$.

Finally, it is easy to see that the identities (3.11) and (3.12) still apply, and lead to

$$\lim_{\lambda\to 0} \int_{0-}^{\infty} \frac{1}{\lambda+w}\,\sigma(dw) = \infty \ .$$

This completes the proof.

It remains to eliminate the two hypotheses as described in the Introduction, and this is also straightforward.

COROLLARY 3.2. The class of all Lévy measures of inverse local times of gap diffusion is the same as that of Theorem 3.1.

PROOF. The definition of the processes (in natural scale) carries over from Theorem 1.1 with obvious modifications (we use B instead of B^+, and there are two values $c^+ > 0$ and $c^- < 0$, with B_{c^-,c^+} killed at both c^+ and c^-). Then it is easy to see that the proof of Theorem 2.1 still applies (except that $m_0 = 0$ is possible if and only if

$$0 \in \operatorname{supp}((I_{(-\infty,0)}(x) + I_{(0,\infty)}(x))m(dx))) \ .$$

This justifies assuming $m_0 = 1$. Then in the persistent case we can write, as in [4, §6.1],

$$\ell^{(-1)}(t) = t + \ell_+^{(-1)}(t) + \ell_-^{(-1)}(t) \quad ,$$

where $\ell_+^{(-1)}$ and $\ell_-^{(-1)}$ are as described in (1.6) with $m_0 = 0$, $c^{-1} = 0$. It follows from Theorem 3.1 that there are Lévy measures

$n_+(dy)$ and $n_-(dy)$ satisfying those conditions, and the Levy measure of $\ell^{(-1)}(t)$ is obviously $n(dy) = n_+(dy) + n_-(dy)$. But the conditions are closed under linear combination, so the same class is obtained. It is also clear from the construction of [4, §6.1] that the same decomposition is valid in the nonpersistent case. One obtains b^+ and b^- much as before, and Theorem 2.2 shows that one can assume $b^+ = b^- = 0$ (by deleting the excursions of one sign, we may work separately with the process of positive excursions and the process of negative excursions). Thus the general exponent is again

$$m\{0\}\lambda + (b^+ + b^-) + \int_0^\infty (1 - e^{-\lambda y})(n_+(dy) + n_-(dy)) \quad ,$$

and the same class $\{n(dy)\}$ is obtained. The reader may satisfy himself as to the details of this generalization.

References

1. H. DYM and H.P. McKEAN. *Gaussian Processes, Function Theory, and the Inverse Spectral Problem.* Academic Press, New York, 1976.

2. W. FELLER. Generalized second order differential operators and their lateral conditions. *Illinois J. Math. 1* (1957), 459-504.

3. H. GZYL. Lévy systems for time-changed processes. *The Annals of Probability 5* (1977), 565-570.

4. K. ITO and H.P. McKEAN Jr. *Diffusion processes and their sample paths.* Academic Press, New York, 1965.

5. I.S. KAC and M.G. KREIN. On the spectral function of the string. *Amer. Math. Society Translations 2, Vol. 103* (1974), 19-102.

6. J.F.C. KINGMAN. Homecomings of Markov processes. *Adv. Appl. Probability 4* (1973), 66-102.

7. F.B. KNIGHT and A.O. PITTENGER. Excision of a strong Markov process. *Z. Wahrscheinlichkeitstheorie verw. Gebiete 23* (1972), 114-120.

8. F.B. KNIGHT. *Essentials of Brownian Motion and Diffusion.* Mathematical Surveys No. 18. American Mathematical Society, Providence, 1981.

9. M.G. KREIN and NUDEL'MAN. *The Markov Moment Problem and Extremal Problems.* Translations of math. monographs, Vol. 50. American Mathematical Society, Providence, 1977.

10. U. KÜCHLER. Some asymptotic properties of the transition densities of one-dimensional quasi-diffusions. *Publ. R.I.M.S. Kyoto Univ. 16* (1980), 245-268.

11. S. WATANABE. On time inversion of one-dimensional diffusion processes. *Z. Wahrscheinlichkeitstheorie verw. Gebiete 31* (1975), 115-124.

12. D. WILLIAMS. *Diffusions, Markov Processes, and Martingales, Vol. 1.* J. Wiley and Sons, London, 1979.

Frank B. Knight
Department of Mathematics
University of Illinois
Urbana, IL 61801, U.S.A.

LEVY SYSTEMS AND PATH DECOMPOSITIONS*

by

J.W. PITMAN

1. Introduction

Itô [21] introduced the idea of a point process attached to a Markov process X, and subsequent work of Weil [42], Getoor [11], [12] and Maisonneuve [29] has shown that the existence of a suitably Markovian Lévy system for such a point process can be instrumental in establishing path decompositions of the Markov process. A *path decomposition,* or *splitting time theorem,* is a result to the effect that some fragment of the trajectory of X is conditionally independent of some other fragment given suitable conditioning variables, usually with one or more of the fragments being conditionally Markovian. Millar [32] gives a survey of such results, and more recent work may be found in the papers of Getoor, Pittenger, and Sharpe: [12], [14], [15], [16], [17], [18], [36], [37], [40]. Lévy systems suitable for deriving path decompositions were constructed in varying degrees of generality by Watanabe [41] and Benveniste and Jacod [2] for the point process of jumps, and by Itô [21], Dynkin [10] and Maisonneuve [28] for point processes of excursions.

The purpose of this paper is to explain in terms of point

*Research supported in part by NSF Grant 78-25301

processes exactly how a Lévy system induces a path decomposition, and hopefully to convince the reader by examples that this is the "right" way to think about a great many splitting time theorems. When viewed in terms of a suitable point process Π of excursions, these splitting time theorems amount to a decomposition of Π at the first time $t = \tau_\omega$ that one of the points $\pi_{t\omega}$ hits a set $A_{t\omega}$, which may in general depend either optionally or predictably on information up to time t. Loosely stated, this first hit decomposition for the point process Π declares that a regular conditional distribution for π_τ given pre-τ information is Q_τ conditioned on A_τ, where $Q_{t\omega}$ is the kernel in a Levy system for Π. The intuitive basis for this decomposition is the obvious splitting of the information that $\tau = t$ into past and present components

(1.1) $$\{\tau=t\} = \{\pi_s \notin A_s,\ 0 < s < t\} \cap \{\pi_t \in A_t\} ,$$

where in Markovian applications "present" will usually mean "future" because π_t will carry in it information about all or part of the future $\theta_t = (X_{t+s},\ s\geq 0)$, and for such non adapted point processes Π some constraint is required to ensure that the event $\{\pi_s \notin A_s,\ 0 < s < t\}$ really depends only on the past at time t (cf. Williams [43], III-79). The algebraic splitting (1.1) could be used directly to prove an analogous decomposition in discrete time, with just a sequence of random variables (π_t) instead of a point process, and with $Q_{t\omega}$ the conditional probability distribution of π_t given pre-t information instead of a Lévy kernel. But for continuous time point processes, (1.1) typically amounts to an identity of null sets, and the calculus of Lévy systems is required to integrate this slippery information.

When attached to a Markov process, the point process Π may be, as in Maisonneuve [28], the process of futures (θ_t) restricted to t in some random countable set, such as the set of times the process exits from a point. Then the point process decomposition at a *first* hit transforms into a path decomposition at a *last* exit. To see how this happens, realize that the last time you leave a point is the first time you never return there in the future. So indeed, "the last shall be first, and the first last".

The application of Lévy systems to prove path decompositions seems to have been initiated by Weil [42], who showed that for certain terminal times τ there is conditional independence of the strict past $(X_t, 0 \leq t < \tau)$ and the future $(X_{\tau+s}, s \geq 0)$ given the left limit $X_{\tau-}$ at time τ. An extension of this result will be presented in Section 3 as an application of the general point process decomposition. In essence, the proof of the point process decomposition in Section 2 is just a reworking of Weil's argument in greater generality with appropriate attention to detail.

In Section 4 the point process decomposition is applied to the exit system of Maisonneuve [29] for excursions away from a homogeneous closed optional set M. The result obtained here is a decomposition at the last time τ in M before a stopping time T. As will be shown in the examples of Section 4, 5, and 6, this includes both the well known last exit decompositions of Pittenger and Shih [38] and Getoor and Sharpe [13], and more recent splitting time theorems of Maisonneuve [29] and Jeulin and Yor [25]. This generalized last exit decomposition is further refined by Getoor and Sharpe [18] in this volume.

Evidently, still more splitting time theorems may be obtained by applying the generalized last exit decomposition in the manner of Millar [33] and Getoor [12] to last exits of some process derived from X rather than last exits of X itself, for example to obtain decompositions at minima, but such applications will not be undertaken here.

2. Decomposition of a Point Process at a First Hitting Time

Throughout this section let $\Pi = (\pi_{t\omega}, t \in D_\omega, \omega \in \Omega)$ be an S-valued point process with domain D_ω, a countable subset of $(0,\infty)$ for each $\omega \in \Omega$. Here $(\Omega,\mathcal{F},P)$ is a complete probability space and $(S,\mathcal{S})$ is a measurable space. With the value of $\pi_{t\omega}$ taken to be a dead point $\partial \notin S$ for $t \notin D_\omega$, it is assumed that Π is a product measurable map from $(0,\infty)\times\Omega$ to $(S_\partial,\mathcal{S}_\partial)$, where $S_\partial = S \cup \{\partial\}$ and $\mathcal{S}_\partial$ is generated by $\mathcal{S}$ and $\{\partial\}$. Associated with Π are the counting processes $(N_{Bt}, t\geq 0)$ which give for $B \in \mathcal{S}$ the number of points N_{Bt} in B up to time t:

$$(2.1) \qquad N_{Bt\omega} = \sum_{s\leq t} 1(\pi_{s\omega} \in B), \qquad t > 0,$$

where a sum such as this can always be restricted to s in the countable domain D_ω of Π. For background on such point processes see Itô [21], Meyer [30], and section III-I of Jacod [22]. It will be assumed for simplicity that Π is *σ-discrete.* That is to say, there exists a sequence of sets S_n with union S such that for each n the process Π is *discrete on* S_n, meaning that the set $\{t: \pi_{t\omega} \in S_n\}$ is almost surely a discrete subset of $(0,\infty)$.

Let $(\mathcal{F}_t, t\geq 0)$ be a filtration in $\mathcal{F}$. The optional and predictable σ-fields on $(0,\infty) \times \Omega$, defined relative to $(\mathcal{F}_t)$, will be denoted by $\mathcal{O}$ and $\mathcal{P}$ respectively. It is *not* assumed that Π is optional except where indicated, and this generality will be important in applications to Markov processes.

(2.2) DEFINITION. A *predictable target* is a set $A \in \mathcal{P} \times \mathcal{S}$, viewed as the map $(t,\omega) \to A_{t\omega}$, where $A_{t\omega} \in \mathcal{S}$ is the section of A at time $t > 0$ and $\omega \in \Omega$.

Think of the S-measurable subset $A_{t\omega}$ of S as a randomly moving target for the point $\pi_{t\omega}$. Intuitively, a predictable target is a *predictable S-valued stochastic process* but it seems useless to try and formalize this by the conventional means of imposing a measurable structure on S.

(2.3) DEFINITION. For a predictable target A, define the *debut* D_A of A by

$$D_A(\omega) = \inf\{t: \ \pi_{t\omega} \in A_{t\omega}\}, \qquad \omega \in \Omega$$

and the *first hit* F_A *of* A by

$$F_A(\omega) = \begin{cases} D_A(\omega) & \text{if the inf defining } D_A(\omega) \text{ is attained,} \\ \infty & \text{otherwise.} \end{cases}$$

Because F is assumed P-complete, each of the times D_A and F_A is a *random time*, that is, an F-measurable random variable with values in $[0,\infty]$. (See Dellacherie [9]). Of course the distinction between D_A and F_A may be neglected if Π is discrete on S, but it certainly cannot when Π is only σ-discrete, as is the case for most interesting point processes attached to Markov processes.

Call a random time τ a *point time* if

$$\tau_\omega \in D_\omega \quad \text{whenever} \quad \tau_\omega < \infty.$$

Then $\tau = F_A$ is a point time such that $\pi_\tau \in A_\tau$ on $\{\tau<\infty\}$, where e.g. $\pi_\tau(\omega) = \pi_{\tau(\omega),\omega}$. Interest in first hits of predictable targets is generated by the following theorem. Its proof will be given as (2.15) below.

(2.4) THEOREM. If (F_t) is the filtration generated by Π, every point time that is a stopping time is the first hit of some predictable target.

(2.5) DEFINITION. A *predictable Lévy system* for Π is a pair (L,Q) consisting of predictable random measure $dL_{t\omega}$ on $(0,\infty)$ and a predictable kernel $Q_{t\omega}(ds)$ from $((0,\infty)\times\Omega,\ P)$ to $(S,\mathcal{S})$, such that for all positive $P \times \mathcal{S}$ measurable $f = f_{t\omega}(x)$, $t > 0$, $\omega \in \Omega$, $x \in S$,

$$\underset{\omega}{E} \sum_t f_{t\omega}(\pi_{t\omega}) = \underset{\omega}{E} \sum_t Q_{t\omega}(f_{t\omega})\, dL_{t\omega}.$$

By considering f of the form $f_{t\omega}(x) = Z_{t\omega}\, 1(x \in B)$ for predictable Z and using a standard monotone class argument, condition (2.3a) amounts to the property that for each $B \in \mathcal{S}$ the counting process

$$(N_{Bt},\ t \geq 0)$$

has for dual predictable projection (or compensator) the process

$$\left(\int_0^t Q_s(B)\, dL_s, \quad t \geq 0 \right).$$

See Dellacherie [9]. Assuming that $(S,\mathcal{S})$ is a nice enough measurable space, a Lévy system for an arbitrary σ-discrete point process Π may be obtained by a disintegration of the Doléans measure $E\Sigma$ on $P \times \mathcal{S}$ as in the proof of Theorem (3.11) in Jacod [22].

A good example to keep in mind is the case when (F_t) is generated by a homogeneous Poisson point process Π with characteristic measure Q, as in Itô [21]. Then N_{Bt} is a Poisson random variable with mean $Q(B)t$, and a predictable Lévy system (L,Q) is provided by the (non-random) Lebesgue measure L, $dL_t = dt$, and the kernel Q that is constantly the characteristic measure Q.

Recall that for an arbitrary random time τ, the *strict pre-*τ *σ-field* $F_{\tau-}$ on $\{\tau<\infty\}$ is defined by declaring that Y is $F_{\tau-}$ measurable iff $Y = Z_\tau$ for some predictable process Z.

(2.6) THEOREM. *(First hit decomposition of a point process).* Let Π be a point process with predictable Lévy system (L,Q). Let $\tau = F_A$ be the first hit of a predictable target A with debut D_A, and assume there exists a stopping time T such that

a) $D_A \le T$ and $\pi_t \notin A_t$ for $D_A < t \le T$ almost surely on $\{D_A<\infty\}$.

Then,

b) $E\, Z_\tau 1(\tau < \infty, \pi_\tau \in B) = E \int_0^T Z_t\, Q_t(BA_t)\, dL_t$ for all $B \in S$ and positive predictable processes Z,

c) $0 < Q_\tau(A_\tau) < \infty$ a.s. on $\{\tau<\infty\}$,

d) $Q_\tau | A_\tau$ serves, where defined, as a regular conditional distribution for π_τ given $F_{\tau-}$ on $\{\tau<\infty\}$: for $B \in S$

$$P(\pi_\tau \in B \mid F_{\tau-}) = Q_\tau(B \mid A_\tau) \quad \text{a.s. on } \{\tau<\infty\},$$

where for a measure Q on (S,S), $C \in S$, $Q|C$ is the probability measure Q conditional on C defined if $0 < Q(C) < \infty$ by $Q(B|C) = Q(BC)/Q(C)$, $B \in S$.

REMARK. Condition a) says that if the target A is hit at all, the stopping time T comes at or after the début of A, but strictly before any further points hit the target. So, where it is finite, $\tau = F_A$ is both the *first hit of* A and *the last hit of* A *up to and including time* T. If Π is optional this condition is trivially satisfied by $T = D_A$. The rôle of the condition is apparent from the first step of the proof.

PROOF. The argument is a familiar one, due originally to Weil [42] in the special case discussed in the next section, and repeated frequently under various disguises in the literature of last exits and excursions: see for example Getoor and Sharpe [13], [17], [18], Getoor [11], and Maisonneuve [28], [29]. Take

$$f_{t\omega}(x) = Z_{t\omega}\ 1\{x \in BA_{t\omega},\quad t \le T_\omega\}$$

in (2.5a) to obtain the identity b) after noticing that a) ensures there is a.s. at most one positive term in the sum of (2.5), namely the term at $t = D_A(\omega)$, and that even this term is zero if the début is not a hit. The remaining assertions are now obtained by repeated application of the identity b). Let $q_{t\omega} = Q_{t\omega}(A_{t\omega})$, which is predictable by Fubini's theorem. Substituting $B = S$ and $Z = 1$ in b) shows that

e) $$L_\omega\{t:\ t \le T_\omega,\ q_{t\omega} = \infty\} = 0 \quad \text{a.s.},$$

and putting first $Z = 1\{q = \infty\}$ and then $Z = 1\{q = 0\}$ in b) yields c). Turning to d), take $F \in \mathcal{F}_{\tau-}$ with $F \subset \{\tau<\infty\}$, say $F = Z_\tau\{\tau<\infty\}$ where $Z \in \mathcal{P}$. Put

f) $$Y_{t\omega} = Z_{t\omega}\ Q_{t\omega}(B|A_{t\omega})$$

with the convention that $Q_{t\omega}(B|A_{t\omega}) = 0$ if $Q_{t\omega}(A_{t\omega})$ is 0 or ∞. Then $Y \in \mathcal{P}$, and

g) $\quad L\{t:\ t \le T,\ Y_t\, Q_t(A_t) \ne Z_t\, Q_t(BA_t)\} = 0$ a.s.

Thus,

$$P(F, \pi_\tau \in B) = E \int_0^T Z_t\, Q_t(BA_t)\, dL_t$$

$$= E \int_0^T Y_t\, Q_t(A_t)\, dL_t \qquad \text{by g)}$$

$$= E\, Y_\tau \{\tau < \infty\} \qquad \text{by b)}$$

$$= E\, F\, Q_\tau(B|A_\tau) \qquad \text{by f).}$$

Since $Q_\tau(B|A_\tau)$ is evidently $F_{\tau-}$ measurable, the conclusion d) is immediate.

(2.7) EXAMPLE. Suppose τ is a predictable point time. Then obviously $\tau = F_A$ for the predictable target A defined by

$$A_{t\omega} = \begin{cases} \emptyset, & t < T_\omega \\ S, & t \ge T_\omega. \end{cases}$$

In this instance (2.6d) reduces to formula (3.25) of Jacod [22].

(2.8) EXAMPLE. Suppose Π is (F_t)-optional and $H \in S$ is such that Π is discrete on H. Let τ_n be the n-th time t that $\pi_t \in H$, and put $\tau_0 = 0$: then taking

$$A_{t\omega} = H\, 1(\tau_{n-1}(\omega) < t)$$

makes τ in the theorem equal to τ_n, and one finds that (2.6d) holds in this instance with $A_\tau = H$. For the case of a Poisson process with characteristic measure Q, this amounts to the result of Itô [21] that the points $\pi_{\tau_1}, \pi_{\tau_2}, \ldots$ are independent and identically distributed with law $Q|H$, independent also of the times $\tau_1, \tau_2, \ldots$. The case when Π is discrete on S and $H = S$ may be found in Brémaud [3], VIII T2.

In applications of the decomposition to point processes attached to Markov processes, the Lévy system kernel Q is usually of the form

$$Q_{t\omega} = Q^{X^-(t,\omega)} \tag{2.9a}$$

where X is a Markov process assumed to have a left limit process X^- in a state space $(E,\mathcal{E})$, and Q is a kernel from $(E,\mathcal{E})$ to $(S,\mathcal{S})$. Typically one wants to apply the decomposition to a target set A of the form

$$A_{t\omega} = \begin{cases} A^{Y(t,\omega)} & \text{for } (t,\omega) \in V, \\ \emptyset & \text{otherwise.} \end{cases} \tag{2.9b}$$

where Y is a predictable process with some other space $(\bar{E},\bar{\mathcal{E}})$, A^y is the section at y of a product measurable subset of $\bar{E}\times S$, and V is a predictable set. The following Corollary amounts to a reformulation of the decomposition by a change of variable which may be applied when a) and b) hold for two arbitrary predictable processes X^- and Y. In the statement it is assumed for simplicity that the special case obtains with

$$(E,\mathcal{E}) = (\bar{E},\bar{\mathcal{E}}) \quad \text{and} \quad X^- = Y,$$

but the general case is obtained by applying the special case with (X^-,Y) instead of Y. Define a measure m on $(E,\mathcal{E})$ by

$$(2.10)\qquad m(C) = E\int_0^T 1(Y_t \in C)\, dL_t, \qquad C \in \mathcal{E}.$$

(2.11) COROLLARY. With the above assumptions in addition to those of (2.4):

a) $P(\tau < \infty,\ Y_\tau \in dy,\ \pi_\tau \in B\ ) = m(dy)\ Q^y(A^y B), \qquad B \in \mathcal{S}.$

b) $P(\tau < \infty,\ \ Y_\tau \in \{y\colon Q^y(A^y) = 0 \text{ or } \infty\}) = 0$

c) $\mathcal{F}_{\tau-}$ and π_τ are conditionally independent given Y_τ on $(\tau<\infty)$, and the conditional distribution of π_τ given $\mathcal{F}_{\tau-}$, $\tau<\infty$, and $Y_\tau = y$, is $Q^y|A^y$.

There is also the following result, which is of interest only if Π is not optional:

(2.12) *Optional form of the decomposition:* Replace "predictable" by "optional" in all of the above definitions and hypotheses, and the conclusions remain valid with $\mathcal{F}_\tau$ instead of $\mathcal{F}_{\tau-}$. Also an alternative to condition (2.6a) in this case is that there exist a stopping time T such that

a) $\qquad D_A < T$ and $\pi_t \in A_t$ for $D_A<t<T$ almost surely on $D_A<\infty$.

PROOF. Just the same.

It only remains to prove (2.4). Here are two preliminaries. Their proofs are elementary, and omitted.

(2.13) LEMMA. Let G be a sub σ-field of F, $Y: \Omega \to S$ an F/S measurable map. Then the σ-field $\sigma(G,Y)$ generated by G and Y consists of all events F in F of the form

$$F = \{\omega: \; Y_\omega \in H_\omega \},$$

where H_ω is the section at ω of a set $H \in G \times S$.

(2.14) LEMMA. Let T be a random time, (F_t) a filtration. Let $H \subset (T < \infty) \times S$. Then $H \in F_{T-} \times S$ iff there exists $A \in P \times S$ such that for every $\omega \in \Omega$ the section of H at ω is the section of A at (T_ω, ω).

(2.15) PROOF OF (2.4). Let B_n be a sequence of sets with union S such that Π is discrete on B_n. Put

$$\tau_n = \begin{cases} \tau & \text{if } \pi_\tau \in B_n \\ \infty & \text{otherwise.} \end{cases}$$

If A_n represents τ_n then the union of the A_n represents τ, so it suffices to consider the special case when for some $B \in S$ with Π discrete on B the time τ satisfies

$$\pi_\tau \in B \quad \text{on } \{\tau < \infty\}. \tag{2.16}$$

Now assume (2.16) and redefine

$$\tau_n = \inf\{t: \; N_{Bt} = n \},$$

so that,

$$\tau\, 1\{\tau<\infty\} = \sum_n \tau_n\, 1\{\tau=\tau_n\} ,$$

where $\{\tau=\tau_n\} \in F_{\tau_n}$. It is a straightforward matter to verify, using the simple structure of discrete point process histories as described in (3.39b) of Jacod [22] or appendix A2 of Brémaud [3], that

$$F_{\tau_n} = \sigma(F_{\tau_n -}, \pi_{\tau_n}). \tag{2.17}$$

Now the preliminaries above imply the existence of $A^n \in P\times S$ such that $\{\tau=\tau_n\} = \{\pi_{\tau_n} \in A^n_{\tau_n}\}$, whence $\tau_\omega = \inf\{t>\tau_{n-1}: \pi_{t\omega} \in A^n_{t\omega}\}$ on $\{\tau=\tau_n\}$. One finally obtains the desired representation by defining

$$A_{t\omega} = \begin{cases} A^n_{t\omega} & \text{for } \tau_{n-1}(\omega) < t \le \tau_n(\omega) \\ \emptyset & \text{otherwise.} \end{cases}$$

3. An Extension of Weil's Decomposition

Let $(\Omega, F, F_t, X_t, \theta_t, P^x)$ be a Ray process with state space (E,E) and consider the E valued point process

$$X^J = (X_t,\ t \in J_\omega,\ \omega \in \Omega)$$

where $J_\omega = \{t>0:\ X_t \neq X_{t-}\}$ is the random set of jump times of X. Following work of Watanabe [41], Benveniste and Jacod [2] established the existence of a Lévy system (L,Q) for X^J with L a predictable additive functional and Q of the form

$$Q_{t\omega} = N^{X(t-,\omega)}$$

where N is a kernel on E. An easy adaptation of (2.15) shows that every stopping time τ such that $\tau \in J$ if $\tau < \infty$ may be represented as the first hit by X^J of some predictable target $A \in \mathcal{P} \times \mathcal{E}$. The conclusion of the point process decomposition (2.6) is that

(3.1) *the conditional distribution of* X_τ *given* $\mathcal{F}_{\tau-}$ *is equal to* $N^{X(\tau-)}|A_\tau$ *on* $\{\tau<\infty\}$.

Taking $A_{t\omega} = H^{X(t-,\omega)}$ where H^x is the section at x of $H \in \mathcal{E} \times \mathcal{E}$, the conclusion of (2.11c) is that, for an arbitrary predictable set V, if $\tau_{H,V}(\omega) = \inf\{t>0: (t,\omega) \in V, (X_{t-},X_t)(\omega) \in H\}$ and if $\tau = \tau_{H,V} + \infty\{\text{inf unattained}\}$,

(3.2) *the conditional distribution of* X_τ *given* $\mathcal{F}_{\tau-}$ *is equal to* $N^x|H^x$ *on* $\{\tau < \infty, X_{\tau-} = x\}$.

For $V = (0,\infty) \times \Omega$ this is the formula of Weil [42].

To illustrate the identity (2.11a) in this case, consider a Lévy process X in $\mathbb{R}^n$ with Lévy measure μ. Then $L(dt) = dt$ and N^x is the measure μ shifted by x. Letting $m^x_{H,V}(dy)$ be the *pre-*$\tau_{H,V}$ *occupation measure*, the expected amount of time spent by the process in dy before $\tau_{H,V}$ starting at x, one finds that

$$(3.3) \qquad P^x(\ \tau<\infty,\ X_{\tau-}\in dy,\ X_\tau \in B\ ) = m^x_{H,V}(dy)\ \mu(BH_y-y).$$

Closely related identities for Markov chains may be found in Pitman [34].

As noted by Weil, the formula (3.2) and the strong Markov property at the stopping time τ imply a decomposition of the process into fragments $(X_t,\ 0\le t<\tau)$ and $(X_{\tau+s},\ 0\le s<\infty)$ which are conditionally independent given $X_{\tau-}$ on $\{\tau<\infty\}$. More generally, for τ as in (3.1), one immediately obtains

(3.4) THEOREM. *The conditional distribution of* θ_τ *given* $F_{\tau-}$ *equals* $\tilde{P}^{X(\tau-)}|(X_0\in A_\tau)$ *on* $\{\tau<\infty\}$, *where* $\tilde{P}^x = \int N^x(dy)P^y$.

In the terminology of Getoor and Sharpe [16], X has a left Markov property at time $\tau = \tau_{H,V}$, a result which does not seem to be included in their survey except in the terminal time case due to Weil.

For comparison with the results of the next section, it is instructive to view the above path decomposition as an instance of the general point process decomposition applied to the point process of futures

$$\Theta^J = (\theta_t,\ t\in J_\omega,\ \omega\in\Omega),$$

for which it is easy to see that the pair $(L,\tilde{P}^{X(t-)})$ serves as a predictable Lévy system. It is Maisonneuve's construction of similar Lévy systems for the process of futures Θ^G on suitable non-optional random sets G that allows an analogous derivation of last exit decompositions.

4. A Generalized Last Exit Decomposition

Let $(\Omega,F,F_t,X_t,\theta_t,P^x)$ be the canonical right continuous realization of a strong Markov process with state space E. Following Maisonneuve [28], let M be a homogeneous closed optional subset of $(0,\infty)\times\Omega$, such as the closure in $(0,\infty)$ of $\{t>0\colon X_t\in B\}$, where B is a Borel subset of E. Let G be the random *exit set* of left ends of intervals contiguous to M (excluding zero), and call a random time τ an *exit time* if $\tau\in G$ on $\{\tau<\infty\}$. For $t\ge 0$ define

$$G_t = \sup M(0,t], \qquad g_t = \sup M(0,t).$$

Let T be a random time. Each of the random times

$$\tau_+(T) = G_T\{0<G_T<T\} + \infty\{G_T = 0 \text{ or } T\}$$

and

$$\tau_-(T) = g_T\{0<g_T<T\} + \infty\{g_T = 0 \text{ or } T\}$$

is an exit time with some right to be called the last exit before T. ($\tau_+ = \tau_-$ except if T is the right end of a contiguous interval, in which case τ_- is the left end of that interval but $\tau_+ = \infty$). To handle both cases, call τ a *last exit time before* T if

(4.1) $$\tau = \tau_-(T) \text{ on } \{\tau<\infty\}.$$

Consider now the *exit process*

$$\Theta^G = (\theta_{t\omega},\ t \in G_\omega,\ \omega \in \Omega)$$

as a point process with values in (Ω, F^*), where F^* is the universal completion of $\sigma(X_t,\ t \geq 0)$. Assuming that X is a right process, Maisonneuve [28] established the existence of an *optional exit system* $(L,\hat{P})$, consisting of a random measure L on $(0,\infty)$ and a kernel $\hat{P}$ from (E,E^*) to (Ω,F^*) such that $(L,\hat{P}^{X(s)})$ is an optional Lévy system for the exit process Θ^G.

Applying the optional form (2.12) of the point process decomposition to the exit process, one immediately obtains

(4.2) THEOREM. *(Generalized last exit decomposition).* Let τ be a last exit time before a stopping time T, and suppose that

$$\tau_\omega = \inf\{t: \ t \in G_\omega, \ \theta_{t\omega} \in A_{t\omega}\}$$

for an optional target $A \in \mathcal{O} \times \mathcal{F}^*$. Then $\hat{P}^{X(\tau)}|A_\tau$ is a.s. well defined on $\{\tau<\infty\}$, and serves under each P^x as a regular conditional distribution for θ_τ given $\mathcal{F}_\tau$.

NOTE. The assumption that τ is an exit time forces the infimum to be attained, so (2.12a) applies.

Of course, the optional forms of (2.6b) and (2.11) may be applied here as well. In particular, if as in (2.9) one has $A_{t\omega} = A(U_{t\omega})$ for some optional process U, the conclusion is that $\mathcal{F}_\tau$ *and* θ_τ *are conditionally independent given* X_τ *and* U_τ. Using the terminology of Getoor [12], τ is then a *splitting time with auxiliary variable* U_τ.

An easy adaptation of (2.15) shows that an arbitrary exit time τ may be represented as the début and first hit of some predictable target set A, so the hypothesis in (4.2) that $\tau = D_A$ for an optional target A is certainly no restriction. In practice the target A to match $\tau = \tau_-(T)$ or $\tau = \tau_+(T)$ for a stopping time T can be obtained more constructively in the following manner, which is well illustrated by the examples below. One starts by writing down for each $t > 0$ the decomposition of Courrège and Priouret [8].

$$(4.3) \qquad T_\omega = t + T^{t\omega}(\theta_{t\omega}) \quad \text{for } T_\omega > t,$$

where each $T^{t\omega}$ is a stopping time, and in general there may be exceptional null sets due to the Markovian completions. But if (4.3) holds exactly for all $t \in G_\omega$ and the map $(t,\omega,\omega') \to T^{t\omega}_{\omega'}$ is $\mathcal{O} \times F^*$ measurable, it is easy to see that for $\tau = \tau^-(T)$ one may take

$$A_{t\omega} = \begin{cases} \{\omega' : T^{t\omega}_{\omega'} \le R_{\omega'}\} & \text{for } t < T(\omega), \\ \emptyset & \text{otherwise,} \end{cases}$$

where $R = \inf M$, and replacing $\{T^{t\omega} \le R\}$ by $\{T^{t\omega} < R\}$ gives an A for $\tau^+(T)$. As shown by Getoor and Sharpe [18] in this volume, this construction can be made to work for an arbitrary stopping time T, provided due allowance is made for completions. In the same paper Getoor and Sharpe show how Theorem (4.2) can be applied under stronger hypotheses to obtain the conditional law of the excursion straddling T given its starting point, its length, its finishing point and information outside the excursion interval. Notice that for $\tau = \tau_+(T)$ the question of whether the (τ,T) fragment $(X_{\tau+s}, 0<s<T-\tau)$ has a Markov property conditional on F_τ on $\{\tau<\infty\}$ is reduced by the construction based on (4.3) to whether the σ-finite Maisonneuve laws $\hat{P}^x$, which make the process $(X_s, 0<s<\infty)$ Markovian with the original semigroup, make $(X_s, 0<s<T^\tau)$ Markovian conditional on $\{T^\tau < R\}$, given the stopping time T^τ (i.e. which stopping time it is, not its value!). In particular it is clear that the (τ,T) fragment has a homogeneous Markov property if T^τ almost surely selects a terminal time (see (5.6) for an example), and this condition might well be necessary too - cf. Sharpe [40].

Another question of interest is that of giving a more explicit description of the pre-τ process. For progress in this direction see Meyer, Smythe and Walsh [31], Getoor and Sharpe [15] and Propositions (3.10) and (3.13) of Jeulin and Yor [25].

(4.4) EXAMPLE. Assume Ω is equipped in the usual way with killing operators k_t, let $R = \inf M(0,\infty)$ and for $t \in G_\omega$ let $e_t = k_R \circ \theta_t$ be the excursion starting at t. Let $B = (B_{t\omega})$ be an optional target set, and define σ_B and τ_B to be the début and first hit of B for the excursion process $(e_t, t \in G_\omega, \omega \in \Omega)$. Let $T_B = \inf M[\sigma_B,\infty)$, which is the right end of the first excursion $e_{t\omega} \in B_{t\omega}$, if there is one, and the unattained infimum σ_B otherwise. Then T_B is a stopping time, $\tau_B = \tau_-(T_B)$ is a last exit before T_B, and Theorem (4.2) applies with

$$A_{t\omega} = \begin{cases} k_R^{-1}(B_{t\omega}) & \text{for } t < T_B(\omega), \\ \emptyset & \text{otherwise}. \end{cases}$$

(Presumably $\tau = \tau_B$ is just about the most general last exit before a stopping time, but it would seem to be necessary to stop the excursions instead of killing them to achieve this, and some further regularity of M might be required.) The case when $B_{t\omega}$ is a constant set $H \in F^*$ is particularly important (cf. Example (2.8)). Subscripting by H instead of $B = (0,\infty)\times\Omega\times H$ in this instance, τ_H is the start of the first excursion of type H, if there is such an excursion, and $\tau_H = \infty$ otherwise. Clearly τ_H is a splitting time without any auxilliary variables, and as in (2.11c) the conditional law of the future θ_τ given F_τ and $X_\tau = x$ on $\{\tau<\infty\}$ is the Maisonneuve exit law $\hat{P}^x$ conditioned on $\{k_R \in H\}$. Taking $H = (\zeta=\infty)$ where ζ is the lifetime gives results of Pittenger and Shih [37] and Meyer, Smythe and Walsh [31], noted already as a consequence of the Maisonneuve theory by Getoor [12]. Further details of the decomposition for $H = (\zeta>t)$ are described in Maisonneuve [28] and Getoor [11]. The case $H = (\zeta>T)$ for an exact terminal time T is also of special interest, since τ is then the start of the excursion straddling T. In this instance (4.3) holds with $T^{t\omega} \equiv T$ and the (τ,T) fragment conditional on F_τ is Markov with semigroup that of the original process killed at T conditional on $(T < R)$, with only the entrance law depending on the value of X_τ. (cf. Getoor and Sharpe [17])

(4.5) EXAMPLE. Let $\alpha_t = t - G_t$, the *age* of the excursion straddling t, let β be a positive optional process, and consider the stopping time

$$T = \inf\{t: \ \alpha_t > \beta_{G_t} \},$$

and the last exit $\tau = \tau_+(T)$. Clearly one may apply (4.2) with

$$A_{t\omega} = \begin{cases} \{\omega': \ R(\omega') > \beta_{t\omega} \}, & t < T_\omega \\ \emptyset & t \geq T_\omega , \end{cases}$$

to obtain the conclusion of (4.2) with

$$A_\tau(\omega) = \{\omega': \ R(\omega') > \beta_\tau(\omega) \} \qquad \text{on } \{\tau<\infty\},$$

where obviously $\beta_\tau(\omega) = \alpha_T(\omega) = T(\omega)-\tau(\omega) > 0$ if $\tau(\omega) < \infty$. Thus τ is a splitting time with auxilliary variable $\beta_\tau = \alpha_T$ on $\{\tau<\infty\}$. It will be argued in Section 6 that this is just a reformulation of Theorem 2 of Maisonneuve [29]. According to Theorem (8.1) of Maisonneuve [28], the (τ,T) fragment is in this instance an inhomogeneous Markov process with transition probabilities depending only on the value of α_T. Note the special case $\beta_t = (u-t)^+$ for a constant time u, which corresponds to $T = u$, and the case $\beta_t \equiv u$, when τ is the start of the first excursion whose length exceeds u. A variety of conclusions to be drawn from the decomposition in these cases may be found in the papers of Getoor and Sharpe [11], [13], [17], and Maisonneuve [28].

(4.6) THE PREDICTABLE VERSION. There is of course a predictable version of Theorem (4.2), assuming the existence of a predictable exit system. It appears necessary to assume either that $T \notin G$ a.s., or that M has no isolated points: compare conditions (2.12a) and (2.6a) to see why.

Maisonneuve [29] constructs a predictable system for the exits of a Hunt process from a set M satisfying a technical condition. An important case when there exists a predictable exit system for an arbitrary strong Markov process is the case when M is the closure of $\{t: X_t=b\}$ for a single point $b \in E$. Ignoring the trivial situation when M is discrete, the reader can easily verify that an exit system is provided in this case by $(L,\hat{P})$, where L is a local time at b and $\hat{P}$ is a constant kernel, that is a single measure, which may be described in terms of the Ito excursion law Q of the next section by the intuitively obvious factorization.

$$(4.7) \qquad \hat{P}(\, k_R \in d\omega,\ R \in dt,\ \theta_R \in d\omega' \,) = Q(d\omega,\ \zeta \in dt)\ P^b(d\omega')\ .$$

5. Decompositions at Exits from a Point

Consider in this section the situation when M is the closure of the set of times t that the strong Markov process X visits a non-absorbing point $b \in E$. To focus on the interesting case assume $P^b(R=0) = 1$, so M has no isolated points a.s. As shown by Itô [21],

Meyer [30] and Maisonneuve [26], [27], there then exists a local time process L for M, whose right continuous inverse $(\tau_s, s \geq 0)$ is a subordinator, and if the excursion process

$$(e_t, t \in G_\omega, \omega \in \Omega)$$

is time changed to form a new point process

$$(\mathcal{e}_{s\omega}, s \in D_\omega, \omega \in \Omega)$$

defined by

$$\mathcal{e}_{s\omega} = e_{\tau_{s-}}(\omega), \qquad D_\omega = \{s: \tau_{s-}(\omega) < \tau_s(\omega)\},$$

this new process is a homogeneous Poisson point process with intensity measure of the form $dt\, Q(d\omega')$ on $(0,\infty)\times\Omega$, where Q is a σ-finite measure on $(\Omega,\mathcal{F})$, the *Itô excursion law,* and the Poisson process is absorbed at the first excursion of infinite length in the transient case. For an elementary proof of these facts, see Greenwood and Pitman [19].

The Poisson property of the time changed excursions may be expressed thus (see Jacod [22], (3.34))

(5.1) *an* $(\mathcal{F}_{\tau_s}, s\geq 0)$*-predictable Lévy system for* $(\mathcal{e}_{s\omega})$ *is* (l,Q)*, where* $dl_{s\omega}=ds$*, and* $Q_{s\omega}$ *is identically equal to the Itô excursion law.*

By inverting the time change this becomes

(5.2) *an* $(\mathcal{F}_t)$*-predictable Lévy system for* $(e_{t\omega})$ *is* (L,Q)*, where* L *is the local time.*

Thus Theorem (2.6) may be immediately applied to the point process $(\pi_t) = (e_t)$, and the condition (2.6a) is superfluous since one can always take $T = \inf M[D_A,\infty)$.

(5.4) EXAMPLE. Let H be a fixed set of excursions with $0 < Q(H) < \infty$,

$$\tau = \inf\{t:\ t \in G_\omega,\ e_t \in H\},$$

the start of the first excursion of type H. Then $P^b(\tau<\infty) > 0$ and the excursion e_τ is *independent* of $F_{\tau-}$ on $\{\tau<\infty\}$ with distribution $Q|H$. (cf. (2.8), (4.3). For some applications see Greenwood and Pitman [20].)

(5.5) EXAMPLE. Specializing Example (4.4) to this case with $\beta_t = (u-t)^+$ so $\tau = \tau_+(u)$, the last exit before u, let m^x be the measure on $(0,\infty)$ defined by $m^x(0,t] = E^x L_t$. Then (2.11a) yields, for $B \in F$,

a) $$P^x(\tau\in dt,\ e_\tau\in B) = m^x(dt)\, Q(B\{\zeta > u-t\}) \quad \text{on} \quad (0,t),$$

from which a host of formulas to be found in Getoor [11] and Chung [7] can be generalized to the present context. Notice that the case of Markov chains (Chung [4], [5], [6]) which provided analogies and motivation for Chung's work in the Brownian case is *included* in the present discussion by virtue of the Ray-Knight compactification. (See Williams [44] III.88). In particular, one immediately obtains from a) the last exit decomposition of the semigroup (P_t) of X

b) $$P_u(x,H) = Q_u(x,H) + \int_0^t q_t(H)m^x(dt), \qquad b \notin H \in E,$$

where (Q_t) is the semigroup of Q, that is the original semigroup killed at b, and (q_t) the entrance law of Q: $q_t(H) = Q(X_t \epsilon H)$. And for $\rho = \inf M(u,\infty)$ the end of the excursion straddling u which started at τ,

$$\text{c)} \qquad P^x(\tau \epsilon dt,\ \rho > r) = m^x(dt)\,\mu(r-t,\infty),$$

where μ is the Q distribution of ζ. It is amusing to note, through some obvious changes of variable, that μ is just the Lévy measure of the subordinator $(\tau_s,\ s \geq 0)$ inverse to L, m^x is the total occupation measure of (τ_s) when X starts at x, and $\tau = \tau_{\sigma-}$, $\rho = \tau_\sigma$ where $\sigma = \inf\{t\colon \tau_s \geq u\}$, so the formula c) is actually an example of the more general Lévy process formula (3.3).

(5.6) EXAMPLE. This is a spatial variant of Example (4.4), and could easily be set up in the generality of that example. The purpose is to show how the decomposition for Brownian motion given in Theorem 3.7 of Jeulin and Yor [25] can be generalized to arbitrary Markov processes, something which is not at all obvious from their method of grossissement developed by Barlow [1] and Jeulin and Yor [23], [24].

Let $f\colon E \to [0,\infty)$ be measurable with $f(b) = 0$, let $(U_t,\ t \geq 0)$ be a positive predictable process, and consider the stopping time

$$T = \inf\{t\colon\ f(X_t) > U_t\}$$

and the last exit before T, $\tau = \tau_+(T)$. Let $A(u) = \{T_u < \infty\}$, where T_u is the terminal time $T_u = \inf\{t\colon\ f(X_t) > u\}$. Then, on the event of probability one where M has no isolated points,

$$\text{a)} \qquad \tau = \inf\{t\colon\ t \in G(0,T],\ e_t \in A(U_t)\}.$$

Applying (2.6) one finds that

b) *A conditional distribution for* e_τ *given* $F_{\tau-}$ *on* $\{\tau<\infty\}$ *is* $Q|A(U_\tau)$.

But under $Q|A(u)$ the process $(X_t,\ 0<t<T_u)$ is Markov with the transition semigroup $(K_t^u,\ t \geq 0)$ of X conditioned to hit $\{\zeta>u\}$ before b and killed on reaching $\{\zeta>u\}$. Thus.

c) *Conditional on* $F_{\tau-}$ *on* $\{\tau<\infty\}$ *with* $U_\tau = u$, *the process* $(X_{\tau+s},\ 0<s<T-\tau)$ *is Markov with semigroup* $(K_t^u,\ t \geq 0)$.

Taking X to be Brownian motion and $f(x) = |x|$, $b = 0$, this is a case of the Jeulin-Yor result, since the work of Williams [43] identifies the modulus of the conditioned process with a three dimensional Bessel process started at 0 and killed on first reaching u. Jeulin and Yor actually consider the Brownian case with two separately moving barriers but the present method adapts easily to show that their formulae in that case may be extended to an arbitrary one dimensional diffusion. For further information about excursion laws of one dimensional diffusions see Rogers [39] and Pitman and Yor [35].

6. Maisonneuve's Filtration

Let M be a random closed optional set relative to a filtration (F_t). Let

$$G_t = \sup M[0,t], \qquad D_t = \inf M(t,\infty), \qquad \alpha_t = t - G_t.$$

In the excursion theory setting of section 4, Maisonneuve [28], [29] studied stopping times of the filtration $(\check{F}_t)$ defined by $\check{F}_t = F_{G_t}$.

To show that the result of Example (4.5) above is just another way of expressing Theorem 2 of [29], requires first the following representation theorem. Its proof will be given as (6.6) below.

(6.1) THEOREM. Let T be an $(\check{F}_t)$ stopping time. Then there exists a positive optional process β such that

$$T = \inf\{t: \alpha_t > \beta_{G_t}\} \quad \text{on } \{G_T<T\}.$$

Secondly, it must be seen that if T is an $(\check{F}_t)$-stopping time then

$$(6.2) \qquad F_{G_T} = \check{F}_T.$$

Maisonneuve noted in [29] the inclusion $F_{G_T} \subset \check{F}_T$, and Getoor and Sharpe [17] made the first step in the other direction by showing that

(6.3) $\check{F}_T$ *and* $F_{G_T} \vee \sigma(\alpha_T)$ *have the same trace on* $\{G_T < T < \infty\}$.

But the representation (6.1) implies that (contrary to a remark of Maisonneuve [29])

$$(6.4) \qquad \alpha_T \text{ is } F_{G_T}\text{-measurable,}$$

and from (6.3) and (6.4) one easily obtains (6.2). That example (4.5) amounts to the same thing as Theorem 2 of [29] is now easily seen using (6.1) and (6.2). Doubtless there is also a predictable version of these results. See section IV of Maisonneuve [29], which contains an error corrected in Section 8 of Getoor and Sharpe [17].

Here is a preliminary for the proof of (6.1), similar to

Proposition (3.40a) of Jacod [22].

(6.5) LEMMA. Let S and T be two $(\check{F}_t)$-stopping times. Then there exists an F_{G_S}-measurable random variable $\hat{T}$ such that

$$T \wedge D_S = \hat{T} \wedge D_S \quad \text{on} \quad \{T \geq G_S\}.$$

PROOF. Fix S. By the usual approximation it suffices to consider the case when T is discrete. Then given a value t of T there exist optional processes Z and Z' such that

$$1\{G_S \leq t = T < D_S\} = Z_t(G_S)\, Z'_t(G_t)\, 1\{G_S \leq t < D_S\} = 1(A_t)\, 1(t < D_S),$$

where

$$1(A_t) = Z_t(G_S)\, Z'_t(G_S)\, 1(G_S \leq t)$$

is F_{G_S}-measurable. But

$$A_t = \{G_S \leq t = \hat{T} < D_S\},$$

where $\hat{T} \in F_{G_S}$ is defined by

$$\hat{T} = t \quad \text{on} \quad A_t \setminus \bigcup_{u<t} A_u ,$$

the union being over the finite number of values u of T with $u < t$. The conclusion follows.

(6.6) PROOF OF (6.1). It suffices to do it for T such that $\alpha_T > \varepsilon$ on $\{T<\infty\}$ for some $\varepsilon > 0$. Indeed, for general T one can then represent T_n defined by

$$T_n = \begin{cases} T & \text{if } \alpha_T > \frac{1}{n} \\ \infty & \text{else} \end{cases}$$

with β_n optional, and $\beta = \inf_n \beta_n$ serves for T. So, supposing $\alpha_T > \varepsilon$ on $T < \infty$, let

$$S_n = \begin{cases} n^{\text{th}} \text{ time } t\colon\ \alpha_t = \varepsilon \\ \infty \quad \text{if no such } t, \end{cases}$$

and apply the lemma to $S = S_n$ to produce $\hat{T}_n \in F_{G(S_n)}$ such that

$$T \wedge D_{S_n} = \hat{T}_n \wedge D_{S_n} \quad \text{on} \quad \{T \geq G_{S_n}\}.$$

Now $\hat{T}_n - G_{S_n} = \beta_n(G_{S_n})$ for some optional β_n. Let

$$\beta = \beta_1[0, D_{S_1}] + \beta_2(D_{S_1}, D_{S_2}] + \ldots \ .$$

Then on $\{G_T < T\}$ it is clear that T is the first $\hat{T}_n < D_{S_n}$, whence on that event

$$T = \inf\{t\colon\ \alpha_t > \beta(t)\} \ .$$

* * * *

I thank Getoor and Sharpe for their helpful criticism of a preliminary draft of this section.

References

1. M. BARLOW. Study of a filtration expanded to include an honest time. *Z. Wahrscheinlichkeitstheorie verw. Gebiete 44* (1978), 307-323.

2. A. BENVENISTE and J. JACOD. Systèmes de Lévy des processus de Markov. *Invent. Math. 21* (1973), 183-198.

3. P. BREMAUD. *Point Processes and Queues: Martingale Dynamics.* Forthcoming book.

4. K.L. CHUNG. On last exit times. *Illinois J. Math. 4* (1960), 629-639.

5. K.L. CHUNG. On the boundary theory for Markov chains II. *Acta. Math. 115* (1966), 111-163.

6. K.L. CHUNG. *Lectures on Boundary Theory for Markov Chains.* Annals of Mathematics Studies Number 65, Princeton University Press, Princeton, 1970.

7. K.L. CHUNG. Excursions in Brownian motion. *Arkiv. for Mat. 14* (1976), 155-177.

8. P. COURREGE et P. PRIOURET. Temps d'arrêt d'une fonction aléatoire: relations d'equivalence associées et propriétés de décomposition. *Publ. Inst. Statist. Univ. Paris 14* (1965), 245-274.

9. C. DELLACHERIE. *Capacités et Processus Stochastiques.* Springer-Verlag, Berlin, 1972.

10. E.B. DYNKIN. Wanderings of a Markov process. *Theo. Prob. Appl. 16* (1971), 401-408.

11. R.K. GETOOR. Excursions of a Markov process. *Ann. Probab. 7* (1979), 244-266.

12. R.K. GETOOR. Splitting times and shift functionals. *Z. Wahrscheinlichkeitstheorie verw. Gebiete 47* (1979), 69-81.

13. R.K. GETOOR and M.J. SHARPE. Last exit decompositions and distributions. *Indiana Univ. Math. J. 23* (1973), 377-404.

14. R.K. GETOOR and M.J. SHARPE. The Markov property at co-optional times. *Z. Wahrscheinlichkeitstheorie verw. Gebiete 48* (1979), 201-211.

15. R.K. GETOOR and M.J. SHARPE. Some random time dilations of a Markov process. *Math. Zeit. 167* (1979), 187-199.

16. R.K. GETOOR and M.J. SHARPE. Markov properties of a Markov process. *Z. Wahrscheinlichkeitstheorie verw. Gebiete 55* (1981), 313-330.

17. R.K. GETOOR and M.J. SHARPE. Excursions of dual processes. To appear.

18. R.K. GETOOR and M.J. SHARPE. Two results in dual excursions. In this volume.

19. P. GREENWOOD and J.W. PITMAN. Construction of local time and Poisson point processes from nested arrays. *J. London Math. Soc. (2) 22* (1980), 182-192.

20. P. GREENWOOD and J.W. PITMAN. Fluctuation identities for Lévy processes and splitting at the maximum. *Adv. Appl. Prob. 12* (1980), 893-902.

21. K. ITO. Poisson point processes attached to Markov processes. *Proc. Sixth Berkeley Symp. Math. Statist. Prob.* pp. 225-239. Univ. of California Press, Berkeley, 1970.

22. J. JACOD. *Calcul Stochastique et Problèmes de Martingales.* Lecture Notes in Math. *714*, Springer-Verlag, Berlin, 1979.

23. T. JEULIN. *Semi-Martingales et Grossissement d'une Filtration.* Lecture Notes in Math. *833*, Springer-Verlag, Berlin, 1980.

24. T. JEULIN et M. YOR. Nouveaux résultats sur le grossissement des tribus. *Ann. Sci. E.N.S. 4e Série, t. 11* (1978), 429-443.

25. T. JEULIN et M. YOR. Sur les distributions de certaines fonctionelles du mouvement brownien. *Séminaire de Probabilités XV (Univ. Strasbourg)*, pp. 210-226. Lecture Notes Math. *850* Springer-Verlag, Berlin, 1981.

26. B. MAISONNEUVE. Ensembles régénératifs, temps locaux et subordinateurs. *Séminaire de Probabilités V (Univ. Strasbourg)*, pp. 147-169. Lecture Notes Math. *191*. Springer-Verlag, Berlin, 1971.

27. B. MAISONNEUVE. Systèmes régénératifs. *Astérisque*, No. 15, *Soc. Math. France*, Paris, 1974.

28. B. MAISONNEUVE. Exit systems. *Ann. Prob. 3* (1975), 399-411.

29. B. MAISONNEUVE. On the structure of certain excursions of a Markov process. *Z. Wahrshceinlichkeitstheorie verw. Gebiete 47* (1979), 61-67.

30. P.A. MEYER. Processus de Poisson ponctuels, d'aprés K. Itô. *Séminaire de Probabilités V (Univ. Strasbourg)*, pp. 177-190. Lecture Notes Math. *191*, Springer-Verlag, Berlin, 1971.

31. P.A. MEYER, R.T. SMYTHE, and J.B. WALSH. Birth and death of Markov processes. *Proc. Sixth Berkeley Symp. Math. Statist. Probab. vol. 3*, pp. 295-306. Univ. California Press, Berkeley, 1972.

32. P.W. MILLAR. Random times and decomposition theorems. *Proc. Symp. Pure Math. vol. 31*, Providence, 1976.

33. P.W. MILLAR. A path decomposition for Markov processes. *Ann. Prob. 6* (1978), 345-348.

34. J.W. PITMAN. Occupation measures for Markov chains. *Adv. Appl. Prob. 9* (1977), 69-86.

35. J.W. PITMAN and M. YOR. A decomposition of Bessel bridges. To appear.

36. A.O. PITTENGER. Regular birth times for Markov processes. *Ann. Prob.* To appear.

37. A.O. PITTENGER and M.J. SHARPE. Regular birth and death times. *Z. Wahrscheinlichkeitstheorie verw. Gebiete*, to appear.

38. A.O. PITTENGER and C.T. SHIH. Coterminal familes and the strong Markov property. *Trans. Amer. Math. Soc. 182* (1973), 1-42.

39. L.C.G. ROGERS. Williams' characterization of the Brownian excursion law: proof and applications. *Séminaire de Probabilités XV (Univ. Strasbourg)*, pp. 227-250. Lecture Notes Math. *850*. Springer-Verlag, Berlin, 1981.

40. M.J. SHARPE. Killing times for Markov processes. To appear.

41. S. WATANABE. On discontinuous additive functionals and Lévy measures of a Markov process. *Japan J. Math. 34* (1964), 53-79.

42. M. WEIL. Conditionnement par rapport au passé strict. *Séminaire de Probabilités V (Univ. Strasbourg)*, pp. 362-372. Lecture Notes Math. *191*. Springer-Verlag, Berlin, 1971.

42. D. WILLIAMS. Path decomposition and continuity of local time for one-dimensional diffusions. *Proc. London Math. Soc. 28* (1974), 738-768.

43. D. WILLIAMS. *Diffusions, Markov Processes, and Martingales; vol. 1: Foundations*. Wiley, New York, 1979.

J. W. PITMAN
Department of Statistics
University of California at Berkeley
Berkeley, CA 94720, U.S.A.

REGULAR BIRTH AND DEATH TIMES FOR MARKOV PROCESSES

by

A. O. PITTENGER

1. Introduction

The concept underlying a homogeneous Markov process is that the evolution of the process from time t onwards depends only on the position at time t and not on the past before t. There are two aspects to this idea. First, there is the conditional independence of the t-past and t-future given the t-present, and second, there is the subsequent evolution of the process as a homogeneous Markov process obeying the same laws of transition. Both aspects are meant by the assertion that the process is Markov or has the Markov property.

During the development of the general theory of Markov processes, it was noticed that for some processes these two features held for a class of non-constant times called optional times or stopping times. These times are characterized by

$$\{T < t\} \in \mathbb{F}_t , \tag{1.1}$$

where $\mathbb{F}_t$ denotes (see Section 3) the usual completion of $\sigma(X_s, s \le t)$, and the property is denoted as the strong Markov property. (For more detailed historical comments see [3] or [14] in which references are given to a number of early papers including those by Doob [4], Blumenthal [2], Hunt [9], and Dynkin and Yushkevich [5].) Of course the definitions of T-past and T-future have to be made precise, but the essential idea is the same as for constant times.

By the symmetry of the definition of conditional independence, it would be reasonable to expect that certain classes of random times would have an analogous property -- but now with the pre-time segment behaving as a homogeneous Markov process. This is indeed the case for terminal times, stopping times T having the additional property that

$$T = t + T \circ \theta_t \qquad \text{a.s. on } \{ T > t \}, \tag{1.2}$$

where θ_t is the usual shift operator. However, now the semigroup describing the evolution of the process up to T is no longer the original semigroup, but rather a conditional version

$$H_t^x(A) = P^x(X_t \in A, t < T \mid 0 < T).$$

These examples lead naturally to the problem of finding other random times with the same features; specifically, times R possessing combinations of the following properties:

(1.3) *The R-past is conditionally independent of the R-future given the R-present.*

(1.4) *The post-R process is a homogeneous Markov process, possibly with different transition probabilities.*

(1.5) *The pre-R process is a homogeneous Markov process, possibly with different transition probabilities.*

When we say a segment of a process is a homogeneous Markov process, we will always mean with respect to the completion of the σ-algebras generated by the segment itself.

The appellation *regular* will denote (1.3), with the proviso that precise definitions of the past, present, and future must be specified and may even have several variants. Times satisfying (1.4) will be called *birth times*, and times satisfying (1.5) will be called *death times*. In this note we will be concerned with *regular birth times* ((1.3) and (1.4)), *regular death times* ((1.3) and (1.5)), and times which are both ((1.3), (1.4) and (1.5)). Characterizations of such times were given in the context of Markov chains by Jacobsen and Pitman and in a general context in papers by Pittenger and by Sharpe. Our purpose here will be to give some of the background relevant to these characterizations, a description of the methods used to obtain the results, and the characterizations themselves. As the informed reader knows, a complete presentation involves a formidable range of technical machinery while a sketchy description fails to do justice to the real need for that machinery. Our approach will be to try to strike a balance between these two extremes and to include sufficient technical detail to illuminate the methods without obscuring the results.

In Section 2 we summarize briefly the history directly leading to these characterizations, with the precise definitions of the process and assorted σ-fields deferred until Section 3, where regular birth times are discussed and the characterization recorded in (3.9). In Section 4 we cover regular death times, the formal statement being given in (4.10), while in Section 5 we examine times that are both regular birth and regular death. Since the derivation of a complete characterization for this last class of times is relatively easy, we include details of the key steps leading up to the final statement in Theorem (5.14).

Before proceeding, we need a word about null sets. The definitions of the random times used here always involve conditions which hold almost surely, and the implied null set may depend on $t \geq 0$. However, under the hypotheses we make, it is always possible to perfect the random times, that is, define equivalent random variables such that, except for a fixed null set, the condition holds for all $x \in E$ and all $t \geq 0$. (See [20] or [21].) To avoid this process, we simply assume the perfected version to begin with.

Finally, we use P to denote both a probability and an expectation so that $E[\ f(X_t(\omega))\cdot 1_\Lambda(\omega)\ ]$ will be written as $P[\ f(X_t),\Lambda\ ]$. This greatly simplifies formulae and should cause no confusion.

2. A Selective History

Rather than attempt a comprehensive discussion of all the papers related to (1.3) - (1.5), we shall instead give a brief account of those directly leading to characterizations of regular birth and death times.

The paper instrumental in initiating this line of research was that of Meyer, Smythe and Walsh [13] in which it was shown that certain classes of non-optional times can be birth times or death times. Following Nagasawa [15] and motivated by time-reversals of processes, hence the prefix co-, they define a time to be *co-optional* if for all $t \geq 0$

$$R\circ\theta_t = (R - t)^+ \quad \text{a.s.} \tag{2.1}$$

(The context of [13] is that of Hunt processes, but the same results hold for right processes.) They then show that R is a death time: the process $\{ X_t, t < R \}$ is a Markov process with semigroup H given by

$$H_t^x(A) = P^x(X_t \in A, t < R \mid 0 < R). \tag{2.2}$$

Note that (H_t) is a new semigroup different from, but absolutely continuous with respect to, the original semigroup.

In (2.1) we used the usual shift operator θ_t whose effect is to make pre-t information unavailable. An analogous operator is k_t, the killing operator, which makes post-t information unavailable:

$$k_t\omega(s) = \begin{cases} \omega(s) & s < t \\ \Delta & s \geq t \end{cases},$$

where Δ is the usual absorbing/killing point in or appended to the state space E. Mimicking the ideas of terminal times, Meyer, Smythe, and Walsh define a *coterminal* time L as a co-optional time with the additional properties

$$L\circ k_s = L \quad \text{on } \{L < s\} \tag{2.3.a}$$

$$L\circ k_s \leq L \quad \text{for all } s. \tag{2.3.b}$$

An additional property necessary for this analysis, but which does not seem to follow from the above, is

(2.3.c) $\quad L \circ k_s$ *is* $\mathbb{F}_s$ *-measurable,*

where $\mathbb{F}_s$ is the usual right-continuous completion of the σ-field generated by $\{ X_r, r \le s \}$. L is called *exact* if $L = \lim [L \circ k_s, s \uparrow \infty]$.

It is shown in [13] that if L is exact, $(X(L+t), t > 0)$ is a homogeneous strong Markov process with semi-group

$$(2.4) \qquad H_t^x(A) = P^x(X_t \in A \mid L = 0);$$

that is, L is a birth time, but with the proviso that $X(L)$ is not included in the post-L process.

Although the emphasis in [13] is on the birth and death properties of the times studied, embedded in the proofs is a certain amount of conditional independence, particularly for the coterminal times. The difficulties are that the L-past defined in [13] contains a little too much information and the probability laws at time L are not defined. These problems were resolved by Pittenger and Shih [18] where the appropriate definition of the L-past, $\mathbb{F}(L)$, and the entrance probabilities at time L were given. Conditional independence was proved, thus showing that L is a regular birth time, but with the observation that the entrance probabilities are only required to be *entrance laws* $(Q_s^x, s > 0)$ with respect to (H_t):

$$(2.5) \qquad Q_{s+t}^x(A) = \int Q_s^x(dy)\, H_t^y(A).$$

The approach used in [18] was quite direct and had the advantage of giving the entrance laws as limits of conditional probabilities of the original process, but the disadvantage of being rather technical and unrelated to more widely used tools. Many of the results of [18] can be obtained in a quite different manner as, for example, in [12]. Millar [14] has a nice discussion of these and related topics.

The next major advance was in the context of discrete time, countable state space Markov chains. Jacobsen and Pitman [11] were the first to pose the general question: characterize all random times with one or more of the properties (1.3), (1.4) and (1.5), in particular regular birth and regular death times. The meaning of "to characterize" is to give algebraic conditions - such as being a terminal time - which correspond to "operational" conditions such as (1.3) and (1.5). (See Jacobsen [10] for a discussion of this phraseology.)

Remarkably enough, given the generality of the question, Jacobsen and Pitman succeeded in finding such a correspondence. Specifically, they defined algebraically two classes of random times $\mathcal{B}$ and $\mathcal{D}$ so that R is regular birth (death) iff R is in $\mathcal{B}(\mathcal{D})$. Moreover, if the chain is irreducible, they also characterized times which are both regular birth and regular death. With the necessary embellishments required by continuous time and general state space, the essence of these characterizations goes over to general Markov processes, (3.9) and (4.10), and that is the theme of the next three sections.

A final comment about [11] is appropriate here. Jacobsen and Pitman discuss times satisfying (1.3), (1.4) and (1.5) separately, obtaining in particular an algebraic characterization of regular times: R is a regular time in the Markov chain case iff R is a splitting time; i.e., for every $n \geq 0$, $\{R = n\} = F_n \cap \theta_n^{-1} G$ where $F_n \in \mathbb{F}$ and $G \in \mathbb{F}$. Regular times have also been analyzed in the general context but as yet the results are not as complete as for regular birth and regular death times. See [8] for the most recent work on regular times.

3. Regular Birth Times

We now need to specify the process. Let X be a right process on a suitable state space E. In particular assume X has a coordinate representation, $X_t(\omega) = \omega(t)$, and that its semigroup (P_t) maps Borel functions into Borel functions. This latter property can be achieved by going to the Ray topology if necessary. Readers unfamiliar with these details may consult [6] or [20], or else simply assume that X is a right continuous strong Markov process with a nice semigroup (P_t) which is normal for $x \in E$: $P^x(X_0 = x) = 1$. We do allow the existence of branching points, but they play no role in the analysis, and all initial measures will be assumed to put zero weight on the set of branching points.

The σ-fields $\mathbb{F}_t$ will denote the usual right-continuous completions of the minimal σ-fields $\mathbb{F}_t^0 = \sigma\{ X_s, s \le t \}$. The *past* of an optional time S is $\mathbb{F}(S) = \{ A\colon A \cap \{S<t\} = A_t \cap \{S<t\}, A_t \in \mathbb{F}_t$ for all $t \}$, and the *past* of an arbitrary random time can be defined as

(3.1) $\mathbb{F}(R)$ *is the* σ-*field generated by* $\mathbb{F}(S) \cap \{S \le R\}$ *as* S *varies over the class of all optional times.*

The foregoing definition was introduced in [18], and there is an equivalent version which is based on concepts from the general theory of stochastic processes. Let $\mathcal{P}$ and $\mathcal{O}$ denote respectively the predictable and optional σ-fields on $[0,\infty) \times \Omega$. $\mathcal{P}$ is generated by left continuous processes and processes evanescent with respect to P^μ, for all μ. $\mathcal{O}$ is defined analogously, but using right continuous processes with left limits. Then it is easy to check [7] that

(3.2) $\mathbb{F}(R)$ *equals the* σ*-field generated by* $Z_R + F1_{\{R=\infty\}}$ *for all* $Z \in \mathcal{O}$ *and* $F \in \mathbb{F}_\infty$.

The characterization of a regular birth time in the general case is given in [16] and is summarized in (3.9) below. However, the definition used there can be weakened somewhat by assuming different measurability conditions on the entrance laws: R is a *regular birth time* if

(3.3.a) there exists a right semigroup (H_t) on a subset F of E and a family of (H_t)-entrance laws $\{ Q_s^x, s > 0 \}$,

(3.3.b) for every $t > 0$ and every bounded Borel measurable f

$$x \to Q_t^x(f) \cdot 1_F(x)$$

is a nearly optional function,

(3.3.c) for every initial law μ and $F \in b\mathbb{F}^0$

$$P^\mu(F\circ\theta_R, R < \infty \mid \mathbb{F}(R)) = Q^{X(R)}(F)\, 1_{\{R<\infty\}} ,$$

where Q^x denotes the probability on path space defined by (Q_s^x) and (H_t).

This definition is from [17] and uses nearly optionality for the measurability condition on $Q^x(f)$ as opposed to the nearly Borel measurability required in [16]. ($g(x)$ is nearly optional if for every ν, $g(X_t)$ is in the σ-field corresponding to $\mathcal{O}$ but defined using only

P^ν-null processes. See [20] for details.) Furthermore, it combines properties (1.3) and (1.4) in one assumption, (3.3.c).

Sharpe has observed that with a minor assumption on $(\Omega, \mathbb{F}^0)$ it is possible to assume conditional independence and Markovness of $(X(R+t), t > 0)$ and then derive the entrance laws rather than assume them. That approach turns out to yield an equivalent definition, and details are given in [17].

Now to the characterization problem. The results of [13] and [18] show that if L is an exact coterminal time, then L is a regular birth time. Since the post-L process is strong Markov, we could construct new regular birth times using such an L. For example, assume ρ: $\mathbb{F} \times \mathbb{F} \to (0,\infty]$ is such that

(3.4.a) $\rho(\omega,\cdot)$ *is optional*

(3.4.b) $\rho(\cdot,\omega')$ *is* $\mathbb{F}(L)$ *measurable.*

Then it is straight forward to show that

(3.5) $$R(\omega) = L(\omega) + \rho(\omega,\theta_L\omega)$$

is also a regular birth time and that the semigroup defined for the post-R process is the same as for the post-L process. This result can be extended to ρ taking values in $[0,\infty]$ provided that

$$\{\ \omega:\ \rho = (\omega,\theta_L\omega) = 0\ \} = \Lambda_0 \cap \theta_L^{-1}\Gamma_0$$

where $\Lambda_0 \in \mathbb{F}(L)$ and $\Gamma_0 \in \mathbb{F}_0$. In this case the post-R semigroup is unchanged but the associated entrance laws are conditional laws of the

L-entrance laws:

$$\tilde{Q}^x(\cdot) = Q^x(\ \cdot \,|\, \Gamma_0\ ).$$

Finally, all of the foregoing could be repeated on an invariant set Γ_1: R equals infinity on Γ_1^c and $L + \rho$ on Γ_1. Again, it is easy to check that the new R is a regular birth time, but now both the semigroup and the entrance laws are obtained by conditioning the original quantities with respect to Γ_1. (This last step is tantamount to introducing a new coterminal time.)

The basic assertion of [16] is that the constructions above exhaust all possible regular birth times. This is a direct extension of Jacobsen and Pitman's result which asserts that in the context of Markov chains every regular birth time is "optional after coterminal", with the invariant set Γ_1 included in the definition of the coterminal time.

Given a regular birth time, the main problem, of course, is the definition of the preceding, coterminal time L, and this rests on a key lemma in [16] involving the absolute continuity of semigroups. Let (H_t) and (P_t) be two semigroups on the same state space, supporting strong Markov processes and with suitable measurability conditions. Then, if H^x and P^x denote the induced probabilities on path space, we have

(3.6) LEMMA. Let $A_0 = \{\ x\colon\ H^x << P^x\ \}$. Then A_0^c is H^x-polar for all $x \in A_0$.

The proof of this is an easy exercise in real variables, and details and precise assumptions are provided in [17].

One can then introduce a right-continuous, non-negative multiplicative supermartingale (M_t) so that M_∞ is a P^x-equivalent version of dH^x/dP^x for $x \in A_0$. The desired coterminal time L is defined as

$$L(\omega) = \sup\{\, t: \ X_t \notin A_0 \ \text{ or } \ \lim_{s \uparrow t} M_{t-s} \circ \theta_s = 0 \,\}.$$

This characterizes L as the last hit of an optional homogeneous set, where a set H in $[0,\infty) \times \Omega$ is homogeneous if its indicator function Z satisfies

$$(3.7) \qquad Z_{s+t}(\omega) = Z_s(\theta_t \omega).$$

It can be shown [7] that L is exact coterminal iff L is the last hit of an optional homogeneous set, and those various definitions are discussed in [17].

The point of (3.6) is that, once the process enters A_0, it stays there under the laws induced by (H_t). L is defined, more or less, so that it is the first coterminal time whose post-L process is consistent with (H_t), and one can use these facts to prove that $L \le R$ a.s. To derive (3.5) with the various constraints on ρ, the approach used in [16] is to discretize R and obtain key measurability properties relating R and L. The technique involves an interplay between the regular birth time properties of L and R and is too technical to reproduce here. It is appropriate, however, to summarize the measurability results, which will be used in section 5:

$$(3.8.a) \qquad \{\, R < t \,\} \in \mathbb{F}_t \cap \{\, L < t \,\},$$

(3.8.b) $\{ R < L+t \} \in \mathbb{F}(L + t),$

(3.8.c) There is an optional ρ such that $R = \rho$ a.s. on $\{L = 0\}$.

Finally, all of the above is recorded in [16] as

(3.9) THEOREM. R is a regular birth time iff there exists a coterminal time $L \leq R$, for each initial measure μ a function ρ satisfying (3.4), and sets Λ_0, Γ_0, and Γ_1 as described after (3.5), so that P^μ-almost surely

$$R(\omega) = \begin{cases} L(\omega) + \rho(\omega, \theta_L \omega) & \text{if } \omega \in \Gamma_1, \\ +\infty & \text{otherwise.} \end{cases}$$

4. Regular Death Times

Even as regular birth times turn out to be optional after coterminal, so regular death times are co-optional before terminal. Jacobsen and Pitman [11] give this in the Markov chain case, and Sharpe [19] has the proof for the general case, which is summarized in (4.10) below. Formally, R is *co-optional before a terminal time* T if it satisfies the algebraic conditions

(4.1.a) $R \leq T$ *a.s.*,

(4.1.b) $R \circ \theta_S = (R - S)^+$ *a.s. on* $\{S < T\}$ *for all optional* S.

This definition also requires T to be a *right* terminal time, which means the set of T-regular points is suitably measurable. Sharpe requires this for his analysis, and we will implicitly assume all terminal times are right. The requirement that (4.1.b) holds for optional S allows the usual perfection results to be obtained. If R is co-optional before ζ, then (4.1.b) need be assumed only for constant times.

To see that (4.1) is sufficient for R to be a death time, we give the following simple calculation based on [13]:

$$(4.2)\qquad \begin{aligned} P[\, f(X_s)g(X_{s+t}),\ s+t < R\,] &= P[\, f(X_s),\ s < T,\ \theta_s^{-1}\{g(X_t), t<R\}\,] \\ &= P[\, f(X_s),\ s < T,\ H_t(X_s,g)\ P^{X_s}(0<R)] \\ &= P[\, f(X_s),\ s < R,\ H_t(X_s,g)\,], \end{aligned}$$

where we have used

$$(4.3)\qquad H_t(x,g) = H_t^x(g) = P^x[\, g(X_t),\ t < R \mid 0 < R\,].$$

The foregoing calculation is valid for optional S as well as constant s, thus proving that (H_t) is a semigroup and the process

$$(4.4)\qquad Y_t = \begin{cases} X_t & \text{on } \{\, t < R\,\} \\ \Delta & \text{on } \{\, t \geq R\,\} \end{cases}$$

is a strong Markov process with state space

$$(4.5)\qquad C_R = \{\, x:\ c(x) \equiv P^x(R > 0) > 0\,\}\,.$$

(One also has to have C_R measurable - i.e. $1_{C_R}(X)$ is in $\mathcal{O}$ - in order to guarantee suitable measurability properties for the process.)

Before describing how (4.1) suffices for R to be regular, we need to make the terms in (1.3) precise. The R-past will now be $\mathbb{F}(R-)$, which can be defined as in (3.1) but with $S < R$, or as in (3.2) but with $Z \in \mathcal{P}$ instead of $Z \in \mathcal{O}$. The R-future is still essentially $\sigma\{ X(R+t), t \geq 0 \}$ plus the null sets. (For a discussion of different definitions, see [8].) Finally the present is now construed to be the left germ field $\mathbb{F}_{[R-]}$ which is generated by null sets and functions $f(X_R)- = \lim[f(X(R-t)), t \downarrow 0]$, where f is α-excessive for some $\alpha > 0$. $\mathbb{F}_{[R-]}$ reduces to $\sigma(X(R-))$ if X is a Hunt process, and it corresponds to $\sigma(X(R-1))$ as used by Jacobsen and Pitman.

Getoor and Sharpe [8] give an equivalent definition of $\mathbb{F}_{[R-]}$ as $\sigma(Z_R \cdot 1_{\{0<R<\infty\}}, Z \in \mathcal{H}^g \cap \mathcal{P})$, where $\mathcal{H}^g$ is the σ-algebra on $(0,\infty) \times \Omega$ generated by evanescent processes and by measurable homogeneous processes (see (3.7) for homogeneity) which are left continuous with right limits. In characterizing regular death times, Sharpe relies heavily on these concepts and the interrelationships among $\mathcal{P}$, $\mathcal{O}$, $\mathcal{H}^g$ and $\mathcal{H}^d$, the last σ-field being defined analogously to $\mathcal{H}^g$ but using right continuity with left limits. (See [1].) In fact his operational characterization of conditional independence is based on a theorem from [8]:

(4.6) THEOREM. R is a regular time with the foregoing definitions iff for every $Z \in \mathcal{H}^d$, $0 \leq Z \leq 1$, there exists $\bar{Z} \in \mathcal{P} \cap \mathcal{H}^g$ such that the random measures $Z * A$ and $\bar{Z} * A$ have the same dual predictable projection, where $A = \varepsilon_R \cdot 1_{\{0<R<\infty\}}$.

The proof that (4.1) suffices for R to be regular is then a matter of checking this equivalent version of conditional independence. The argument is short but technical, relying heavily on dual projections and shifts of random measures. We leave it to the reader to check the details ([8; Thm 5.2]).

In proving that regular death times necessarily satisfy (4.1), the first problem, of course, is to define T. Before doing that, however, we give a precise definition of a regular death time R:

(4.7.a) (1.3) holds with the R-past, present, and future as defined above.

(4.7.b) If $c(x) = P^x(R > 0)$, then $c(X)$ is $\mathcal{O}$ measurable.

(4.7.c) The process Y defined in (4.4) is a temporally homogeneous Markov process defined on the state space C_R.

T can be defined using only the Markovian character of the pre-R process. Let (H_t) be its associated semigroup and (P_t) the original semigroup. Once again invoke standard results [3] to produce a non-negative right-continuous multiplicative supermartingale (M_t), so that for $\Lambda \in \mathbb{F}_t$

(4.8) $$H^x(\Lambda) = P^x(M_t, \Lambda)$$

where H^x and P^x are the probabilities on path space induced by the respective semigroups. T is then defined as $\inf\{ t: M_t = 0 \}$, and if $S = \inf\{ t: M_t \notin C_R \}$, it can be shown that $R \le T \le S$ a.s. In fact T is the *smallest* optional time dominating R, [17], a fact that follows easily from (4.9) below.

The proof that $R \le T$ does not require conditional independence, and if one seeks to go as far as possible without (4.7.a), a reasonable goal is to get a more specific expression for M_t. In fact Sharpe shows that

$$(4.9) \qquad M_t \cdot c(X_0) = P^{X_0}(\, R > t \mid \mathbb{F}_t \,),$$

and, if (4.1) held, M_t would be $c(X_t)\cdot 1_{[0,t)}/c(X_0)$. However, this identification does seem to require (4.7.a). Thus, still without (4.7.a), Sharpe proceeds by defining a new multiplicative functional $\bar{M}_t$ as $c(X_0)M_t/c(X_t)$, which "should be" $1_{[0,T)}$, and proves that the dual predictable projection of $\varepsilon_R \cdot 1_{\{0<R<\infty\}}$ is an additive functional with respect to $\bar{M}$. This paves the way for the use of conditional independence via Theorem (4.6) and the resulting proof that R is co-optional for T [19: Thm 3.14].

In summary

(4.10) THEOREM. R is a regular death time (in the sense of (4.7)) iff R is co-optional before a (right) terminal time T (in the sense of (4.1)).

5. Times of Regular Birth and Death

Suppose R is both a regular birth time and a regular death time. How do the characterizations (3.9) and (4.10) interact? In [11] Jacobsen and Pitman show that if the chain is irreducible, R must be either pure terminal or pure coterminal. In general R can be a mixture of these two, as is shown in [17]. Since it is only fitting to prove at least one theorem here, we will provide the key steps leading to this characterization, omitting only some technical details. (For those requiring rigor, these technicalities involve establishing the measurability of L_0 and T_2 below.) To add suspense, the proof will precede the statement of the theorem.

Assume then that R is both regular birth and regular death. We thus have a coterminal L and a terminal T_0 such that:

(5.1) $$L \leq R \leq T_0 \quad \text{a.s.,}$$

and T_0 is the smallest optional time dominating R.

(5.2.a) $$\{ R < t \} \in \mathbb{F}_t \cap \{ L < t \}.$$

(5.2.b) $$\{ R < L+t \} \in \mathbb{F}(L + t) .$$

(5.2.c) *There is an* $(\mathbb{F}_t)$ *optional* ρ *such that* $R = \rho$ *on* $\{L=0\}$.

(5.3) $$R \circ \theta_S = (R - S)^+ \quad \textit{on} \quad \{S < T_0\} \textit{ for all optional } S.$$

From the fact that L is coterminal, it follows that $\varphi(x) = P^x(L > 0)$ is excessive. Defining

(5.4) $$E_0 = \{ x: \ \varphi(x) < 1 \}$$

and using properties of coterminal times, it is easy to check that $(X_{L+t})_{t>0}$ stays in E_0 a.s. on $\{L < \infty\}$. We can partition E_0 into disjoint sets,

(5.5) $A_0 = \{ x: \varphi(x) = 0 \} \qquad A_1 = \{ x: 0 < \varphi(x) < 1 \},$

and note that A_0 is finely closed and absorbing, since it is the null set of an excessive function. If we define

(5.6) $$S_0 = \inf\{ t: X_t \in A_0 \},$$

then $\varphi(X(S_0)) = 0$, and we are ready for the first assertion:

(5.7) LEMMA. $L \le S_0 \le T_0$ a.s.

PROOF. Since

$$P^x(L > S_0) = P^x(L\circ\theta_{S_0} > 0, S_0 < \infty)$$

$$= P^x(\varphi(X(S_0)), S_0 < \infty) = 0,$$

$L \le S_0$ is immediate. For the second inequality, $L \le T_0$ gives

$$0 = P^x(L\circ\theta_{T_0} > 0, T_0 < \infty) = P^x(\varphi(X(T_0)), T_0 < \infty).$$

Hence $\varphi(X(T_0)) = 0$ a.s. on $\{T_0 < \infty\}$, and $S_0 \le T_0$ a.s.

Next we use (5.2.c) and the zero-one law to partition E_0 in a different way:

(5.8)
$$A_i = \{ x: P^x(R > 0 \mid L = 0) = 1 \} \cap E_0$$
$$A_r = \{ x: P^x(R > 0 \mid L = 0) = 0 \} \cap E_0.$$

Thus A_i and A_r are the R-irregular and R-regular points, conditional on $\{L = 0\}$. Assuming enough measurability for A_i and A_r - or circumventing that problem as in [17] - the next result involves the key interplay between the properties of regular birth and regular death

(5.9) LEMMA. For almost every ω, $X_t(\omega) \in A_i$ if $L(\omega) < t < R(\omega)$, and $X_t(\omega) \in A_r$ if $R(\omega) < t < T_0(\omega)$.

PROOF. By (5.2.b), $\{ R > L+t \} \in \mathbb{F}((L+t)+)$. Hence

$$P^x[R > L+t \mid \mathbb{F}((L+t)+)] = 1_{[0,R)}(L+t) \quad \text{a.s.}$$

Using perfection results available from (5.3), we have

$$\{ R > L+t \} = \{ 0 < R\circ\theta_{L+t}, L+t < T_0 \}.$$

Since L + t is a regular birth time,

$$P^x[R > L+t \mid \mathbb{F}((L+t)+)] = P^{X(L+t)}[R > 0 \mid L = 0]\, 1_{[L+t < T_0]} \quad \text{a.s.,}$$

or

$$(5.10) \qquad 1_{[0,R)}(L+t) = 1_{[0,T_0)}(L+t)\cdot 1_{A_i}(X(L+t)) \quad \text{a.s.}$$

Invoking a technical result from [17], both sides of (5.10) are right continuous, hence indistinguishable, and that suffices for (5.9), as can be easily checked. □

We are halfway to the characterization of R. Let $T_2 = T_0 \wedge T_1$, where T_1 is the penetration time of $A_0 \cap A_r$. Since A_0 is absorbing and $L \le S_0 \le T_0$ from (5.7), $X_t \notin A_0$ for $t < L$. Since $X_t \notin A_i$ for $t \in (L,R)$ by (5.9) and $X(L) \in A_r$ forces $R = L$, it follows that $R \le T_2$. But T_0 is the smallest optional time dominating R, (5.1), so that $T_0 \le T_1$. In brief

(5.11) LEMMA. $X_t \notin A_0 \cap A_r$ for all $t < T_0$.

Unfortunately, we don't have a maximality property for L and so must define

(5.12) $$L_0 = L \vee L_1,$$

where L_1 is the last exit time from $A_1 \cap A_i$. Since $L_0 \le S_0$, $L_0 \le T_0$. On $\{R < T_0\}$, $X_t \in A_r$ for all t in (R,T_0) by (5.9), and $S_0 \le T_0$ gives $X_t \in A_0$ for all $t \in [T_0,\infty)$. Hence

(5.13) LEMMA. $L_0 \le R$ and $X_t \notin A_1 \cap A_i$ for $L_0 < t$.

We are finally ready for the complete characterization:

(5.14) THEOREM. A random time is both a regular birth time and a regular death time iff there exist

i) a finely closed absorbing set $A_0 \subset E$

ii) a right terminal time T_0, and

iii) a coterminal time L_0

such that the following are satisfied with S_0 denoting the hitting time of A_0

(5.15) $$L_0 \le S_0 \le T_0 \text{ and } L_0 \le R \le T_0,$$

(5.16) for a.a. ω, the unordered pair $\{ R(\omega), S_0(\omega) \}$ is equal to $\{ L_0(\omega), T_0(\omega) \}$.

PROOF. Suppose R is such a time. Then we have established (5.15). If $R < t < T_0$, (5.9) gives $X_t \in A_r$ and (5.11) gives $X_t \notin A_0 \cap A_r$. Thus $S_0 \geq T_0$, or $S_0 = T_0$ on $\{R < T_0\}$.

If $L_0 < t < R$, then $X_t \in A_i$ from (5.9) and $X_t \notin A_1 \cap A_i$ from (5.13), thus forcing $L_0 = S_0$ on $\{L_0 < R\}$. The combination of these two results gives (5.16) immediately.

To prove the reverse implication, we write R explicitly as co-optional before T_0,

$$R = \begin{cases} T_0 & \text{if } L_0 \vee S_0 < T_0 \\ L_0 & \text{otherwise} \end{cases}$$

and as optional after L_0,

$$R(\omega) = L_0(\omega) + \rho(\omega, \theta_{L_0}\omega),$$

where we use $S'(\omega) = \inf\{ t > 0 : X_t \notin A_0 \}$ in defining

$$\rho(\omega,\omega') = \begin{cases} T_0(\omega') \wedge S'(\omega') & \text{if } L_0(\omega) < T_0(\omega) \\ 0 & \text{if } L_0(\omega) = T_0(\omega). \end{cases}$$

It is easy to check that both representations have their advertised properties, and details are available in [17] for the unbeliever.

References

1. J. AZEMA. Théorie générale des processus et retournement du temps. *Ann. Sci. Ecole Norm. Sup. 4 Série t6* (1973), 459-519.

2. R.M. BLUMENTHAL. An extended Markov property. *Trans. Amer. Math. Soc. 85* (1957), 52-72.

3. R.M. BLUMENTHAL and R.K. GETOOR. *Markov Processes and Potential Theory.* Academic Press, New York, 1968.

4. J.L. DOOB. Markoff chains - denumerable case. *Trans. Amer. Math. Soc. 58* (1945), 455-473.

5. E.B. DYNKIN and A.A. YUSHKEVICH. Strong Markov processes. *Theory Prob. and Appl. 1* (1956), 134-139.

6. R.K. GETOOR. *Markov Processes: Ray Processes and Right Processes.* Lecture Notes in Math. *440.* Springer, Berlin 1975.

7. R.K. GETOOR and M.J. SHARPE. The Markov property at co-optional times. *Z. Wahrscheinlichkeitstheorie verw. Gebiete 48* (1979), 201-211.

8. R.K. GETOOR and M.J. SHARPE. Markov properties of a Markov process. *Z. Wahrscheinlichkeitstheorie verw Gebiete 55* (1981), 313-330.

9. G.A. HUNT. Some theorems concerning Brownian motion. *Trans. Amer. Math. Soc. 81* (1956), 294-319.

10. M. JACOBSEN. Markov chains: birth and death times with conditional independence. Preprint.

11. M. JACOBSEN and J. PITMAN. Birth, death, and conditioning of Markov chains. *Ann. Prob. 5* (1977), 430-450.

12. B. MAISONNEUVE. Exit systems. *Ann. Prob. 3* (1975), 399-411.

13. P.A. MEYER, R. SMYTHE, J.B. WALSH. Birth and death of a Markov process. *Proc. Sixth Berkeley Symp. Math. Stat. Prob., Vol.* III. 295-305. University of California Press, Berkeley, 1972.

14. P.W. MILLAR. Random times and decomposition theorems. *AMS Proc. of Symposia in Pure Math. 31* (1977), 91-103.

15. M. NAGASAWA. Time reversion of Markov processes. *Nagoya Math. J. 24* (1964), 177-204.

16. A.O. PITTENGER. Regular birth times for Markov processes. *Ann. Prob.,* to appear.

17. A.O. PITTENGER and M.J. SHARPE. Regular birth and death times. *Z. Wahrscheinlichkeitstheorie verw. Gebiete,* to appear.

18. A.O. PITTENGER and C.T. SHIH. Coterminal families and the strong Markov property. *Trans. Amer. Math. Soc. 182* (1973), 1-42.

19. M.J. SHARPE. Killing times for Markov processes. *Z. Wahrscheinlichkeitstheorie verw. Gebiete,* to appear.

20. M.J. SHARPE. *General theory of Markov processes.* Forthcoming book.

21. J.B. WALSH. The perfection of multiplicative functionals. Séminaire des Probabilités VI (Univ. de Strasbourg), pp. 233-242. Lecture Notes in Math. *258*. Springer-Verlag, Berlin, 1972.

A.O. PITTENGER
Department of Mathematics
University of Maryland
Baltimore County
Catonsville, Maryland 21228

SOME RESULTS ON ENERGY

by

ZORAN R. POP-STOJANOVIC and MURALI RAO

0. Introduction

The concept of energy we are dealing with here is a generalization of that dealt with in the classical theory. Although rightfully observed by P.A. Meyer in [4] p. 140, that such a generalization "loses delicacy as it gains generality", we will try to add a few results which may somewhat simplify a way in dealing with this concept. To be more precise, in the literature dealing with concept of energy, the basic tools are Dirichlet spaces techniques [7] and the kernel theory, which are both natural offshoots of the classical theory. The symmetry of kernels plays the basic role there.

This paper will try to avoid both the use of Dirichlet spaces and the kernel theory. It will use probabilistic tools such as sub-Markov resolvents, Revuz measures, and additive functionals.

In a large part of this paper the only given elements will be a sub-Markov resolvent and an excessive reference measure. Although we refer to class (D) potentials, this notion does not need a stochastic process in the background. Similarly, D. Revuz used additive functionals to get his measure. However, one can associate a measure without the aid of additive functionals. In short, even though we use these freely - implicitely assuming a nice process in the background, they are not needed. It is done only to simplify the exposition. In the last section however, we do use the process and additive functionals corresponding to excessive functions. The following example will be useful for the last section.

Given a right-continuous potential (X_t). If (X_t) belongs to the class (D), (see [4]), the energy of (X_t) (which may be infinite) is the number $\frac{1}{2}E(A_\infty{}^2)$ denoted by $e(X_t)$, where (A_t) denotes the natural increasing process associated with (X_t). In the case when (X_t) is not in (D) we assume that (X_t) has infinite energy. Furthermore, it can be shown that in the predictable case the following energy formula holds:

$$(*) \qquad e(X_t) = \tfrac{1}{2}E\Big[\int_0^{+\infty} (X_s + X_{s-})\, dA_s \Big],$$

while in the optional case one has

$$e(X_t) = \tfrac{1}{2}E\Big[\int_0^{+\infty} (X_{s+} + X_s)\, dA_s \Big].$$

To illustrate the connection with the classical potential theory, let $U(x,y)$ be the Newtonian kernel in R^3 and μ a positive measure. Then, the Newtonian potential f is $f(\cdot) = (U\mu)(\cdot) = \int U(\cdot,y)\, \mu(dy)$. If (B_t) is a Brownian motion in R^3 with $\Omega = C(R_+,R^3)$, then (X_t), with $X_t = f(B_t)$, is a positive right-continuous supermartingale associated with the potential f. Under relatively mild assumptions such as that μ has a compact support and that it does not charge polar sets in R^3, one can show that f is a potential belonging to the class (D). Then, this implies [5] the existence of the additive continuous functional (A_t) for which $f(B_t)(\alpha) =$

$E^{\alpha}[\, A_\infty - A_t \mid F_t \,]$ and $E^{\bullet}[\, \int_0^{+\infty} f(B_s)\, dA_s \,] = \int U(\cdot,y)\, f(y)\, \mu(dy)$, or $E^{\bullet}[\, \int_0^{\infty} X_s\, dA_s \,] = \int U(\cdot,y)\, f(y)\, \mu(dy)$, where after using continuity of (A_t) one gets:

$$E^{\alpha}[\, A_\infty^2 \,] = 2\, (\alpha, U(f\mu)) = 2\, (f\mu, U\alpha),$$

where the symmetry of U has been used. Along a suitable sequence of α's one gets $\lim_\alpha \frac{1}{2} E^{\alpha}[A_\infty^2] = (\mu, U\mu)$ which gives the relation between the classical energy and that associated with the process (X_t) given by (*).

In the sequel, we shall use notations as in [1].

1. Energy

Let βU^{β}, $\beta > 0$, be a family of sub-Markov resolvants. We assume that hypothesis (L) of P.A. Meyer holds. Let dx denote the excessive reference measure. Let η denote an α-excessive measure. With $(\ ,\ )_\eta$ we will denote the scalar product relative to measure η. When $\eta(dx) = dx$, we will write $(\ ,\ )$.

PROPOSITION 1. Let $\gamma > \alpha > 0$ and suppose $(U^{\gamma}|f|,|f|)_\eta < +\infty$. Then

$$(f,g)_\eta \geq (\gamma - \tfrac{\alpha}{2})(g,g)_\eta, \tag{1}$$

where $g = U^{\gamma} f$.

PROOF. Since βU^{β}, $\beta > 0$, is a sub-Markov resolvent one has

$$(\beta U^{\beta} f)^2 \leq \beta U^{\beta} f^2. \tag{2}$$

After replacing β by $\beta + \alpha$ and f by $g = U^{\gamma} f$ in (2), one gets:

(3) $$((\beta+\alpha)U^{\beta+\alpha}U^{\gamma}f)^2 \le (\beta+\alpha)U^{\beta+\alpha}(U^{\gamma}f)^2.$$

Let us integrate (3) with respect to η and take into account that η is an α-excessive measure. Then,

(4) $$((\beta+\alpha)U^{\beta+\alpha}g,(\beta+\alpha)U^{\beta+\alpha}g)_{\eta} \le \frac{\beta+\alpha}{\beta}(g,g)_{\eta}, \quad \text{with } g = U^{\gamma}f.$$

One can rewrite (4) in the form

(5) $$\frac{\alpha}{\beta}(g,g)_{\eta} + (g-(\beta+\alpha)U^{\beta+\alpha}g,\ g + (\beta+\alpha)U^{\beta+\alpha}g)_{\eta} \ge 0.$$

Since $g = U^{\gamma}f$, the resolvent equation gives $g-(\beta+\alpha)U^{\beta+\alpha}g = U^{\beta+\alpha}f-\gamma U^{\beta+\alpha}g$ and (5) becomes:

(6) $$0 \le \frac{\alpha}{\beta}(g,g)_{\eta} + (U^{\beta+\alpha}f-\gamma U^{\beta+\alpha}g,\ g + (\beta+\alpha)U^{\beta+\alpha}g)_{\eta}.$$

Finally, after multiplying both sides of (6) by $\beta + \alpha$ and letting $\beta \to +\infty$, one gets $0 \le \alpha(g,g)_{\eta} + (f-\gamma g,2g)_{\eta}$, which is (1). □

COROLLARY 1. Let $(U^{\alpha}|f|,|f|)_{\eta} < +\infty$. Then, for $\alpha > 0$,

(7) $$(f,g)_{\eta} \ge \frac{\alpha}{2}(g,g)_{\eta}, \quad \text{where } g = U^{\alpha}f.$$

PROOF. Put $\gamma = \alpha$ in the previous proposition. □

COROLLARY 2. For $f,g \geq 0$, one has

$$U[\, fUf + gUg \,] \geq U[\, gUf + fUg \,]. \tag{8}$$

PROOF. If $U = U^0$ exists then for every x the measure $U(x,dy)$ is α-excessive for all $\alpha \geq 0$. □

REMARK 1. Let $\alpha \geq 0$. If $(U^\alpha|f|,|f|) < +\infty$, one says that $U^\alpha f$ has finite α-energy. Then, the non-negative quantity

$$I(U^\alpha f) = \sqrt{(U^\alpha f,f)} \tag{9}$$

is called the α-energy of $U^\alpha f$.

REMARK 2. From the definition it follows that energy of $U^\alpha f$ is finite iff energy of $U^\alpha|f|$ is finite.

REMARK 3. It is easy to see that if $U^\alpha f$ and $U^\alpha g$ have finite energy so does $U^\alpha f + U^\alpha g$. Indeed, assume $f, g \geq 0$. It is sufficient to show that for each $\alpha > 0$,

$$(U^\alpha f,g) + (U^\alpha g,f) \leq (U^\alpha f,f) + (U^\alpha g,g). \tag{10}$$

If, in addition, f and g are square integrable all quantities in (10) are finite. Using (7) with $\eta = dx$ one concludes that $(U^\alpha(f-g), f-g\,) \geq 0$, i.e., that (10) holds. The transition from square integrable functions to arbitrary non-negative functions follows by usual limiting procedures.

REMARK 4. For $\alpha > 0$, it follows from (7) that if $U^{\alpha}f$ has finite energy then $U^{\alpha}f$ is square integrable. Also, the energy of $U^{\alpha}f$ is zero only if $U^{\alpha}f = 0$. For $\alpha = 0$ this question remains open. Thus, at least for $\alpha > 0$ energy induces a proper norm. This norm will be denoted by $\| \ \|_e$.

2. Separability

A natural question arises: to characterise separability with respect to this norm. Answering this question one obtains the following theorem.

THEOREM 1. Let $A = \{ Uf: \ f \geq 0, (Uf,f) < +\infty \}$. Then A is separable with respect to the norm $\|Uf\|_e = (Uf,f)$.

PROOF. This proposition will be proved in several steps.

Step 1. Assume that a sequence Uf_n increases to Uf, where $Uf \in A$. Then Uf_n converges to Uf in energy norm. This is shown exactly as in the classical case.

Step 2. The set $\{ Uf: \ Uf \in A, \ f \in L^1 \}$ is dense in A. Indeed, choose $0 \leq \phi_n \leq 1$ with $f\phi_n$ increasing to f and $f\phi_n \in L^1$. Then, $U(f\phi_n)$ increases to Uf. Finally, use the conclusion from Step 1.

Step 3. The set $\{ Uf: \ Uf \in A, \ f \in L^1 \}$ is separable. Indeed, the set of f such that $f \geq 0$, $f \in L^1$ and $(Uf,f) < +\infty$ is separable (as a subset of L^1). Let f_n be a sequence dense in this set, i.e., for every $f \geq 0$ with $(Uf,f) < +\infty$ one can choose a subsequence f_m such that f_m converges to f in L^1. By choosing a further subsequence if necessary one may assume that f_m converges to f almost everywhere. In particular, the sequence $g_k = \inf_{m \geq k} f_m$ increases to

f almost everywhere. Using Step 1, one concludes that Ug_n converges to Uf in energy. So, one can find k for which $\| Ug_k - Uf \|_e$ is small. Now the sequence $f_k \wedge f_{k+1} \wedge \cdots \wedge f_m = h_m$ decreases to g_k, and it is not difficult to see that Uh_m converges to Ug_k in energy. Thus the set of potentials of finite infima of a sequence dense in L^1 is dense in A. This proves the theorem.

3. Limits of Potentials with Bounded Energy

The purpose of this section is to find the possible limits of potentials Uf_n of a bounded energy. Toward that goal let us introduce a few definitions and properties.

An excessive function s that is finite almost everywhere will be called a class (D) potential if $P_{R_n} s$ decreases to zero almost everywhere as $n \to +\infty$; here

$$R_n = \inf\{ t: \ s(X_t) > n \},$$

where (X_t) is a Markov process along whose paths the behavior of s is being considered.

To every class (D) potential there corresponds a measure, not necessarily σ-finite, called its Revuz measure [6].

Let us also recall that from every sequence of excessive functions one can extract a subsequence which converges almost everywhere to an excessive function. Thus the following simple theorem states that all possible limits of a sequence Uf_n which is bounded in energy are class (D) potentials.

THEOREM 2. Let $s_n = Uf_n$ be a sequence of potentials such that $(Uf_n, f_n) \leq M$, for all n for some $M > 0$. Suppose $\lim_{n\to+\infty} s_n = s$ almost everywhere. Then, s is a class (D) potential.

PROOF. Let $u(\cdot,\cdot)$ be a density of U and let $\phi > 0$ in L^1 be such that

$$\eta(y) = \int u(x,y)\, \phi(x)dx \leq 1. \tag{11}$$

Then $\eta(y)dy$ is excessive. Since $\eta \leq 1$,

$$(Uf_n, f_n\eta) \leq (Uf_n, f_n) \leq M. \tag{12}$$

But

$$(Uf_n, f_n\ \ ) = \int \phi(x)\, U[f_n Uf_n](x)\, dx \tag{13}$$

$$= \tfrac{1}{2} \int \phi(x)\, E^x\Big[\int_0^\infty f_n(X_t)^2\, dt \Big]\, dx.$$

Let E denote the measure

$$\int \phi(x)\, E^x[\quad]\, dx. \tag{14}$$

For almost all x, $s_n(x) < +\infty$ and

$$s_n(X_t) = M_n(t) - A_n(t)$$

where

$$A_n(t) = \int_0^t f_n(X_s)\, ds, \qquad M_n(t) = E^x[\, A_n(\infty) \mid F_t\,]$$

are continuous and the latter makes sense whenever $s_n(x) = E^x[A_n(\infty)]$ is finite. (Family $(\mathcal{F}_t)$ is the right-continuous family of σ-fields associated in the natural way with the process (X_t).) In particular, with respect to measure E, $E[M_n^2(\infty)] = \int \phi(x)\, E^x[M_n^2(\infty)]dx$ is bounded. So, for any stopping time T, $E[M_n^2(T)]$ is also uniformly bounded in n. In particular, $E[s_n^2(X_T)]$ is uniformly bounded and hence the same is true for $E[s^2(X_T)]$. □

In general, the Revuz measures of class (D) potentials are not σ-finite. However, in the case when the approximating potentials are uniformly bounded in energy, this assertion is true:

PROPOSITION 2. If $g>0$ almost everywhere, then $Ug>0$ everywhere.

PROOF. First note that for any non-negative f, f = 0 almost everywhere on the set $\{Uf = 0\}$. Indeed, if B is the set $\{Uf = 0\}$, by the maximum principle $Uf1_B \equiv 0$ and hence $f1_B = 0$ almost everywhere. Thus if $g > 0$ almost everywhere, so is Ug. But $Ug \geq \alpha U^\alpha Ug$ and for each x, $U^\alpha(x,dy)$ is absolutely continuous with respect to dy. Therefore, $U^\alpha Ug$ is positive everywhere. □

THEOREM 3. Let Uf_n increase to s, and for every n, $(Uf_n,f_n) \leq M$, $M > 0$. Then s is finite almost everywhere. Moreover, let μ be the Revuz measure for s. Then μ is σ-finite and

$$(s,\mu) \leq M. \tag{15}$$

PROOF. Let the natural additive functional (A_t) correspond to s. Choose $g>0$ in L^1 such that $p = Ug$ has finite energy. Then,

(16) $$(p,f_n) \le (Ug,f_n) + (g,Uf_n) \le (Uf_n,f_n) + (Ug,g).$$

Now for each N and x such that $s(x) < +\infty$, one has

(17) $$E^x[\int_0^\infty p\wedge N(X_t)\, dA_t\,] = \lim_{n\to\infty} E^x[\int_0^\infty p\wedge N(X_t)\, f_n(X_t)\, dt\,]$$

$$\le \liminf_{n\to\infty} U[f_n Ug](x)$$

Provided that $s < +\infty$ almost everywhere, (17) implies that the potential on the left-hand-side there is less than or equal to the limit inferior of the potentials on the right-hand-side. Hence, the total mass of the Revuz measure of the excessive function on the left-hand-side of (17) is less than or equal to the limit inferior of the total mass on the right-hand-side, i.e.,

(18) $$\mu[p\wedge N] \le \liminf_{n\to\infty} (p,f_n).$$

Using (16) and letting $N \to +\infty$, one gets

$$(p,\mu) \le M + (Ug,g).$$

Since $p > 0$ everywhere, μ must be a σ-finite measure.

Moreover, (16) implies that $(g,Uf_n) \le M + (Ug,g)$. By letting $n \to +\infty$, one obtains $(g,s) < +\infty$. Since $g > 0$, one concludes that $s < +\infty$ almost everywhere as asserted.

After replacing p by Uf_m in (18), one obtains

$$\mu[N\wedge Uf_m] \le \liminf_{n\to\infty} (Uf_m,f_n) \le \liminf_{n\to\infty} (Uf_n,f_n) \le M$$

wherefrom (15) follows after letting $m \to +\infty$, $N \to +\infty$ and using the monotone convergence theorem.

4. Excessive Functions of Finite Energy

A famous theorem of H. Cartan [2] in the classical potential theory asserts that the space of positive measures of finite energy is complete. Here, in a general setting, a complete analogue of this result is being presented.

Let s be excessive. One asserts that s is of finite energy if there exists a constant $M > 0$ such that

$$(s,g) \le M \, \|Ug\|_e = M \, (Ug,g)^{\frac{1}{2}}. \tag{19}$$

Now, let us check that for $s = Uf$ this definition is consistent. First, if Uf is of finite energy then Uf is finite almost everywhere. This fact has been established in Theorem 3. Then, one can find a sequence ϕ_n such that $U(f\phi_n)$ is bounded and $f\phi_n$ is integrable and increases to f as $n \to +\infty$. This is a standard procedure. It follows then that $\| U(f\phi_n)\|_e < +\infty$. Therefore, $(U(f\phi_n),f\phi_n) \le (Uf,f\phi_n) \le M \, \| U(f\phi_n)\|_e$, which implies $\| U(f\phi_n)\|_e \le M$. By letting $n \to +\infty$ one obtains the conclusion as claimed.

THEOREM 4. The completion of the set A defined in Theorem 1 consists of excessive functions of finite energy.

REMARK. One does not assert that all excessive functions of finite energy belong to this completion.

PROOF. Suppose that (Uf_n,f_n) is bounded. Then, (Uf_n,g) is bounded for all g such that $||Ug||_e$ is finite. Since the set A (defined in Theorem 1.) is separable, by choosing a subsequence if necessary, one may assume that

$$\lim_{n\to\infty} (Uf_n,g) \tag{20}$$

exists for every g with $(Ug,g) < +\infty$. Let us show that there exists an excessive function s such that

$$\lim_{n\to\infty} (Uf_n,g) = (s,g)$$

for all g satisfying (20). To do so, fix g so that $(Ug,g) < +\infty$. Then, for every non-negative bounded measurable ϕ, $U(g\phi)$ has finite energy. It follows from (19) that $\lim_{n\to\infty} (Uf_n,g\phi)$ exists. Hence there is a function s_g such that

$$\lim_{n\to\infty} (Uf_n,g\phi) = (s_g,g\phi).$$

If $h \geq g$, $g = h\psi$, one has that

$$\lim_{n\to\infty} (Uf_n,g\phi) = \lim_{n\to\infty} (Uf_n,h\psi\phi) = (s_h,h\psi\phi) = (s_h,g\phi).$$

This shows that $s_h = s_g$, (gdx)-almost surely. Therefore, there is a function s such that

$$\lim_{n\to\infty} (Uf_n,g) = (s,g).$$

Finally, one has to show that s can be chosen to be an excessive function. Indeed, each g such that $Ug < +\infty$ almost everywhere is the limit of an increasing sequence g_m with Ug_m bounded and g_m in L^1. Now, for any ρ, $\rho \in L^1$ and $U\rho \le K$, one has $(U\hat{P}_t\rho, \hat{P}_t\rho) \le K\int\hat{P}_t\rho \le K\int\rho < +\infty$, where $\hat{P}_t$ is a dual semigroup, which always exists. Hence

$$\lim_{n\to\infty} (Uf_n, \hat{P}_t\rho) = (s, \hat{P}_t\rho),$$

implying $(P_t s, \rho) \le (s, \rho)$. The validity of this inequality for all ρ considered means that $P_t s \le s$ almost everywhere. Starting from this, using standard arguments, one can show that s can be chosen excessive. □

5. Convergence in energy

In the previous parts of this paper it has been shown that if a sequence Uf_n is bounded in energy and increases to an excessive function s then s is natural. The obvious question is: if sequence Uf_n is a Cauchy sequence in energy and converges pointwise to s, can one conclude that s is regular in the following sense: whenever a sequence of stopping times T_n increases to T, sequence $P_{T_n} s$ decreases to $P_T s$. Unfortunately, this is not true. To see this, one should start from the fact that for convex sets weak closure and strong closure are the same. Then, Uf_n bounded in energy and converging pointwise to s, would imply that it converges weakly.

However, the statement about regularity is true if the so-called "sector condition" introduced by M. Silverstein in [7] pp. 17, holds.

Methods presented here do not overlap with that of M. Silverstein [7], since neither regularity conditions on the semi-group are imposed nor Dirichlet space methods are used here. Instead, here, the following two simple facts are used.

First: Every natural potential is the sum of bounded potentials.

Second: A sum of regular potentials is regular.

DEFINITION. Let us say that the sector condition is valid if for every *signed* f and g,

$$|(Uf,g)| \leq M\,(Uf,f)^{\frac{1}{2}}(Ug,g)^{\frac{1}{2}}$$

for some constant M, $M > 0$.

THEOREM 5. Suppose the sector condition holds. Then all natural potentials are regular.

PROOF. Let s be a natural potential. One may assume that s is bounded. Let A denote its additive functional. Then, a sequence of potentials of the form Uf increases to s. By taking convex combinations we may assume that there is a sequence Uf_n such that $Uf_n \leq s$ and $\lim_{n\to\infty} Uf_n = s$, where sequence Uf_n is Cauchy in the energy norm. In particular $\lim_{n\to\infty} (Uf_n,f_n)$ exists.

On the other hand, the sector condition implies that

$$\lim_{m,n\to\infty} (Uf_n,f_m) = \lim_{n\to\infty} (Uf_n,f_n). \tag{21}$$

Let μ denote the Revuz measure of s. It is not difficult to show that for each n,

(22) $$\lim_{m\to\infty} (Uf_n, f_m) = (Uf_n, \mu) \le (s, \mu).$$

Using (21) and (22) one concludes that

(23) $$(s,\mu) \ge \lim_{n\to\infty} (Uf_n, f_n).$$

On the other hand, P.A. Meyer has proved in [4] p. 143, that

(24) $$\liminf_{n\to\infty} 2U[f_n Uf_n](x) = \liminf_{n\to\infty} E^x[(\int_0^{+\infty} f_n(X_t)\, dt)^2]$$

$$\ge E^x[A_\infty^2] = E^x[\int_0^{+\infty} [Y_t + s(X_t)]\, dA_t],$$

where $Y_t = (s(X_t))_-$. This allows one to compare the Revuz measures of both sides of the last inequality. Using the facts that $Y_t(\theta_s) = Y_{t+s}$ and $Y_t \ge s(X_t)$, one shows that

(25) $$E^x[\int_0^{+\infty} (Y_t - s(X_t))\, dA_t]$$

is an excessive function. The total mass of the Revuz measure of the excessive function $E^x[\int_0^{+\infty} Y_t\, dA_t]$ is at least that of $E^x[\int_0^{+\infty} s(X_t)\, dA_t]$. However, the last one is equal to (s,μ). Relations (23) and (24) imply that the excessive function in (25) has Revuz measure zero. Hence, the function in (25) is zero, which implies that s is regular.

References

1. R.M. BLUMENTHAL and R.K. GETOOR. *Markov Processes and Potential Theory.* Academic Press, New York, 1968.

2. H. CARTAN. Théorie du potential newtonian: énergie, capacité, suites de potentials. *Bull. Soc. Math. France 73*, 74-106 (1945).

3. L.L. HELMS. *Introduction to Potential Theory.* Wiley-Interscience, New York, 1969.

4. P.A. MEYER. *Probability and Potentials.* Blaisdell, Waltham, 1966.

5. C. DELLACHERIE and P.A. MEYER. *Probabilités et Potentiel, Vol. II.* Hermann, Paris, 1980.

6. D. REVUZ. Measures associeés a fonctionnelles additives de Markov. *Trans. Amer. Math. Soc. 148,* 501-531 (1970).

7. M.L. SILVERSTEIN. The sector condition implies that semipolar sets are quasi-polar. *Z. Wahrscheinlichkeitstheorie verw. Gebiete 41* (1977), 13-33.

Z. POP-STOJANOVIC
Department of Mathematics
University of Florida
Gainesville, FL 32611, U.S.A.

M.K. RAO
Department of Mathematics
University of Florida
Gainesville, FL 32611, U.S.A.

ABSOLUTE CONTINUITY AND THE FINE TOPOLOGY

by

J. WALSH and W. WINKLER

0. Introduction

One of the basic contrasts between the classical and axiomatic theories on the one hand and their probabilistic analogues on the other is that many of the underlying hypotheses of the former are topological, and of the latter, measure-theoretical. A case in point is the regularity of excessive functions, which is assured in the classical and axiomatic settings by assuming lower semi-continuity, and in the probabilistic setting by assuming much weaker conditions such as the absolute continuity condition (hypothesis (L) of Meyer).

The reason for the submergence of topology is that the connection between the Markov process one studies and the topology of the state space is rather weak. There is, however, one topology, the fine topology, which is intrinsically related to the process, and it is our purpose to complete the circle by showing that at least one measure-theoretic condition, hypothesis (L) of Meyer, can be expressed in purely topological terms. We will show that hypothesis (L) holds if and only if the fine topology satisfies the countable chain condition (CC), that every disjoint collection of finely open sets is countable.

We will introduce our notation and definitions in the first

section and prove the main result in the second.

1. Notation and Topological Preliminaries

The basic notation is taken from Walsh and Meyer (1971) (see also Getoor (1975)). Let E be a topological space which is Lusin and metrizable, that is, E can be extended to a compact metric space $\hat{E}$ and E is a Borel subset of $\hat{E}$. Let $\mathcal{E}$ and $\mathcal{E}^*$ denote, respectively, the Borel measurable and the universally measurable subsets of E.

Let $(P_t)_{t\geq 0}$ be a semigroup of Markov kernels on E. If $(P_t)_{t\geq 0}$ is merely sub-Markov, we can adjoin an absorbing point to E in the usual manner and make $(P_t)_{t\geq 0}$ Markov. We suppose that the semigroup transforms the Borel measurable functions into universally measurable functions. We assume that the semigroup satisfies the two "hypothèses droites". The first is:

HD1: *for every law* μ *on* E, *there exists a Markov process whose trajectories are right continuous, whose transition semigroup is* (P_t), *and whose initial law is* μ.

We construct a canonical realization which is right continuous: Ω denotes the set of right continuous functions from $\mathbb{R}_+$ to E, X_t is the coordinate function of index t on Ω, and $\mathcal{F}^o$ and $\mathcal{F}^o_t$ denote the σ-fields generated on Ω by, respectively, $(X_s,\ s \geq 0)$ and $(X_s,\ s \leq t)$ with values in $(E,\mathcal{E})$. We provide Ω with the measures P^μ for which the process (X_t) is Markov, admits (P_t) as its transition semigroup, and μ as its initial law. $\mathcal{F}^\mu$ and $\mathcal{F}^\mu_t$ denote the completions of $\mathcal{F}^o$ and $\mathcal{F}^o_t$ with respect to P^μ. A set $A \subseteq E$ is *nearly Borel* (relative to X) if for each μ there exist Borel subsets B and B' on E such that $B \subseteq A \subseteq B'$ and

$$P^{\mu}(X_t \in B'\backslash B \text{ for some } t \geq 0) = 0.$$

We denote the nearly Borel measurable subsets by $\mathcal{E}^n$.

Hypothesis HD1 implies the right continuity of the semigroup, and one can therefore define the resolvent U_p associated with the semigroup. A nonnegative universally measurable function f is q-*excessive* if $pU_{p+q}f \leq f$ for all $p > 0$ and $\lim_{p\to\infty} pU_{p+q}f = f$ pointwise.

We proceed to the second hypothesis. This hypothesis implies the strong Markov property. It is

HD2: *Let* f *be a* q-*excessive function. Then* f *is nearly Borel measurable, and for* P^{μ}-*almost all* $\omega \in \Omega$, $t \to f(X_t(\omega))$ *is right continuous on* $\mathbb{R}_+$.

For $A \subseteq E$ define, with the usual convention that $\inf \emptyset = \infty$,

$$T_A(\omega) = \inf\{ t > 0: X_t(\omega) \in A \}$$

If $A \in \mathcal{E}^n$, then a point x is *regular* for A if $P^x(T_A = 0) = 1$; otherwise, x is *irregular* for A. If B is an arbitrary set, then a point x is *regular* for B if x is regular for every nearly Borel set A containing B; otherwise, x is *irregular* for B.

A set $N \subseteq E$ is a *fine neighborhood* of x if $x \in N$ and x is irregular for $E\backslash N$. The topology generated by the neighborhoods is the *fine topology* and its members are the *finely open sets*. If $\alpha > 0$, then the fine topology, which we denote by $\mathcal{O}$, has a base consisting of nearly Borel measurable sets.

The two main conditions which will concern us are the countable chain condition on $\mathcal{O}$:

(CC) *Every disjoint collection of finely open sets is countable;*

and the absolute continuity condition of Meyer:

(L) *There exists a finite measure* m *such that* $B \in \mathcal{O} \cap E^*$ *implies that* $m(B) > 0$.

Since the nearly Borel finely open sets form a base for $\mathcal{O}$ we have immediately that (L) ⇒ (CC). In section 2 we will prove that (CC) ⇒ (L).

Before continuing, we need to define two more concepts. Let ν be a finite measure. $A \in E^n$ is called the *fine support* of ν if

i) A is finely closed,

ii) $\nu(E\backslash A) = 0$, and

iii) if C is finely open and $C \cap A \neq \emptyset$, then $\nu(C \cap A) > 0$. A set $A \in E^n$ is *stable* (or *absorbing*) if $P^x($ X hits $E\backslash A) = 0$ for all $x \in A$. A stable set is necessarily finely open.

2. Main Result: (CC) ⇔ (L)

We wish to prove that (CC) ⇔ (L). We have already observed that (L) ⇒ (CC). To show that (CC) ⇒ (L) we prove two lemmas. Throughout the proofs all sets will be assumed to be nearly Borel measurable unless explicitly stated to be otherwise.

LEMMA 1. Let (CC) hold. If μ is a finite measure and if $\nu = \mu U_1$, then there exists $A \in E^n$ such that

a) A is the fine support of ν, and

b) A is stable.

PROOF. If $\nu(C) > 0$ for all $C \in \mathcal{O}$, then $A = E$ and we are done. If not, there exists $C_1 \in \mathcal{O}$ such that $\nu(C_1) = 0$. Put

$$B_1 = \{ x: \ P^x(X \text{ hits } C_1) > 0 \}.$$

Then $C_1 \subseteq B_1 \in \mathcal{O}$, since $B_1 = \{ x: \ E^x[\exp(-\alpha T_{C_1})] > 0 \}$ for any $\alpha > 0$. If $x \in E\backslash B_1$, then $P^x(\text{ X hits } B_1) = 0$. Indeed, if $P^x(\text{ X hits } B_1) > 0$, then there exists a stopping time T such that $P^x(X_T \in B_1) > 0$, and consequently, we have

$$P^x(\text{ X hits } C_1) \geq E^x(\ P(\text{ X hits } C_1 \mid X_T\ )\ ) > 0.$$

Thus, $E\backslash B_1$ is stable. If $\nu(B_1) > 0$, then $E^\mu\{ \int_0^\infty I_{B_1}(X_s)ds \} > 0$, and by the above reasoning, $P^\mu(\text{ X hits } B_1) > 0$ implies that $P^\mu(\text{ X hits } C_1) > 0$ and $\nu(C_1) > 0$. Thus $\nu(B_1) = 0$.

Now we use transfinite induction to construct A. Let $A_0 = E$ and $A_1 = E\backslash B_1$ where B_1 is given above. Then A_1 is stable, B_1 is finely open, and $\nu(B_1) = 0$. Let β be a countable ordinal and suppose $\{A_\alpha\}_{\alpha<\beta}$ has been chosen so that

i) $A_\alpha \supseteq A_{\alpha+1}$ for all $\alpha+1 < \beta$,

ii) $\nu(A_\alpha) = 0$ for all $\alpha < \beta$, and

iii) A_α is stable and finely closed for $\alpha < \beta$.

If for some $\alpha < \beta$ we have that $\nu(A_\alpha \cap C) > 0$ for every $C \in \mathcal{O}$ such that $A_\alpha \cap C \neq \emptyset$, then let $A = A_\alpha$. If not, assume $\beta = \alpha + 1$ and choose $C \in \mathcal{O}$ such that $C \cap A_\alpha \neq \emptyset$ and $\nu(C \cap A_\alpha) = 0$. Set $B_\alpha = \{ x: P^x(\text{ X hits } C \cap A_\alpha\ ) > 0 \}$ and $A_\beta = A_\alpha \backslash B_\alpha$. Then, by the reasoning used in obtaining A_1 and B_1, $\{A_\alpha\}_{\alpha\leq\beta}$ satisfies i), ii) and iii).

If β is a countable limit ordinal, and if i), ii) and iii) hold for all $\alpha < \beta$, set $A_\beta = \bigcap_{\alpha<\beta} A_\alpha$. Then A_β is stable and finely closed. Also

$$\nu(E\setminus A_\beta) \le \sum_{\alpha<\beta} \nu(E\setminus A_\alpha) = 0.$$

Thus, i), ii) and iii) hold for $\{A_\alpha\}_{\alpha\le\beta}$. Let c be the first uncountable ordinal. By (CC) the collection of finely open sets $\{ A_\alpha \setminus A_{\alpha+1}\colon \alpha < c \}$ is necessarily countable. Thus, there exists $\delta < c$ such that $A_\gamma \setminus A_{\gamma+1} = \emptyset$ for $\gamma \ge \delta$.

We note that the above argument goes through with only minor modifications if μ is a countable sum of finite measures.

LEMMA 2. Let (CC) hold. Then there exists a finite measure μ such that μU_1 has fine support E.

PROOF. Let μ_0 be any finite measure. Let $\nu_0 = \mu_0 U_1$ and A_0 be the fine support of ν_0. By Lemma 1, A_0 is finely closed and stable. If $A_0 = E$, we are done. If $A_0 \ne E$, let $\hat{\mu}_1$ be any finite measure such that $\hat{\mu}_1 U_1(E\setminus A_0) > 0$. ($\hat{\mu}_1$ could be δ_x for some $x \in E\setminus A_0$.) Let $\mu_1 = \mu_0 + \hat{\mu}_1$, $\nu_1 = \mu_1 U_1$ and let A_1 be the fine support of ν_1. Then $A_1 \supseteq A_0$ and A_1 is finely closed and stable. Let β be an ordinal and assume that, for each $\alpha < \beta$, μ_α is a finite measure, $\nu_\alpha = \mu_\alpha U_1$, A_α is the fine support of ν_α, and $A_\alpha \subseteq A_{\alpha+1}$ if $\alpha+1 < \beta$. We note that if $A_\alpha = E$ for some $\alpha < \beta$, the transfinite induction would have stopped at that stage. Consequently, we assume that $A_\alpha \ne E$ for $\alpha < \beta$.

Suppose β is of the form $\alpha+1$ where α is a countable ordinal.

Since $A_\alpha \neq E$, we can find a finite measure $\hat{\mu}_\beta$ such that $\hat{\mu}_\beta U_1(E\backslash A_\alpha) > 0$. Let $\mu_\beta = \hat{\mu}_\beta + \mu_\alpha$, let $\nu_\beta = \mu_\beta U_1$, and let A_β be the fine support of ν_β. Then $A_\beta \supseteq A_\alpha$ and $A_\alpha \neq A_\beta$. If β is a countable limit ordinal, let μ_β be a finite measure which is equivalent to $\Sigma_{\alpha<\beta}\, \mu_\alpha$. Let $\nu_\beta = \mu_\beta U_1$ and A_β be the fine support of ν_β. Then $A_\beta \supseteq A_\alpha$ for all $\alpha < \beta$. Now A_α is finely closed and stable. Consequently, A_α and $A_{\alpha+1}\backslash A_\alpha$ are finely open. Since (CC) holds, the collection of finely open sets $\{ A_{\alpha+1}\backslash A_\alpha : \alpha < c \}$ contains at most a countable number of non-empty sets. Thus, there exists $\delta < c$ such that $A_{\gamma+1}\backslash A_\gamma = \emptyset$ for $\gamma \geq \delta$. Then $A_\delta = E$ and μ_δ is the desired finite measure μ.

REFERENCES

1. R.M. BLUMENTHAL and R.K. GETOOR (1968). *Markov Processes and Potential Theory.* Academic Press, New York.

2. R. GETOOR (1975). *Markov Processes: Ray Processes and Right Processes.* Lecture Notes in Mathematics *440.* Springer-Verlag, Berlin.

3. P.A. MEYER (1962). Functionelles multiplicatives et additive de Markov. *Ann. Inst. Fourier, 12*, 125-130.

4. J. WALSH and P.A. MEYER (1971). Quelques applications des resolvante de Ray. *Invent. Math. 14,* 143-166.

J. WALSH
Department of Mathematics
University of British Columbia
Vancouver, B.C. V6T 1W5, CANADA

W. WINKLER
Department of Mathematics
University of Pittsburgh
Pittsburgh, PA 15261, U.S.A.

Since [illegible] we can find [illegible] such that [illegible]

[illegible]

the [illegible] and [illegible]

[illegible] limit optional [illegible] which [illegible]

[illegible]

[illegible] is [illegible] and [illegible]

consequent [illegible]

[illegible] of the hypotheses [illegible] contains at most [illegible]

countable [illegible] thus [illegible]

[illegible] and [illegible] dropped

[illegible]

REFERENCES

1. BLUMENTHAL, R. M. and GETOOR, R. K. (1968). [illegible] Processes [illegible] Academic Press, New York.

2. [illegible]GETOOR (1975). Markov Processes: Ray Processes and Right Processes. Lecture Notes in Mathematics 440. [illegible]

3. [illegible] (1981). [illegible] the [illegible]

4. [illegible]

[illegible] [illegible]
Department of Mathematics Department of Mathematics
University of British Columbia University of Pittsburgh
Vancouver, B.C. V6T 1W5 Canada Pittsburgh, PA [illegible] USA

[5] for examples). Through studies of the latter type we know that every process with stationary and independent increments is the sum of a linear drift term, a constant multiple of a Brownian motion, and a compensated sum of jumps described by a Poisson random measure (due to LEVY and ITO [13]); that every continuous "regular" strong Markov process on $\mathbb{R}$ is obtained from a Brownian motion by a random time change followed by a spatial transformation and killing (due to DYNKIN and his students who added the needed stochastic methods to FELLER's analytic ones, see DYNKIN [6]); that every continuous strong Markov process on $\mathbb{R}^m$ that is also a martingale is obtained from a quasi-diffusion process by a random time change (SKOROKHOD [21]). The present paper extends these by providing representations for all semimartingale Hunt processes on $\mathbb{R}^m$, for all quasi-left-continuous semimartingale additive functionals of Hunt processes on arbitrary spaces, and for some others.

The following is an informal (non-mathematical) account of the main results and issues. We shall give a precise account of all the results in §3, after listing some preliminary definitions and results in §2. For the present, we are concerned with Hunt processes taking values in $\mathbb{R}^m$ and having infinite lifetimes -- these are right-continuous, have left-hand-limits, are strong Markov and quasi-left-continuous, etc. We assume that they are defined over "sufficiently large" probability spaces.

§1a. *Ito processes*

In his fundamental paper [14], ITO introduced a class of Markov processes X on $\mathbb{R}$ that satisfy

$$(1.1)\quad X_t = X_0 + \int_0^t b(X_s)\, ds + \int_0^t c(X_s)\, dW_s$$

$$+ \int_0^t \int_{\mathbb{R}_0} k(X_{s-},z)\, I_{\{|k(X_{s-},z| \le 1\}}[N(ds,dz) - ds \cdot \frac{dz}{z^2}]$$

$$+ \int_0^t \int_{\mathbb{R}_0} k(X_{s-},z)\, I_{\{|k(X_{s-},z)| > 1\}}\, N(ds,dz),$$

where b, c, k are some sufficiently smooth deterministic functions, W is a Wiener process, and N is a Poisson random measure on $\mathbb{R}_+ \times \mathbb{R}_0 = [0,\infty) \times (\mathbb{R}\setminus\{0\})$ with mean measure $n(ds,dz) = ds\, dz/z^2$. The conditions of smoothness on b, c, k in [14] ensure that there is one and only one solution X, and then it is easy to see that X is a Hunt process. In particular, if

$$(1.2)\quad b(x) = b_0,\quad c(x) = c_0,\quad k(x,z) = k_0(z),\quad \int(|k_0(z)|^2 \wedge 1)\frac{dz}{z^2} < \infty,$$

are free of x, then X has stationary and independent increments. Conversely, every process X with stationary and independent increments is obtained in this manner.

Although ITO [14] assumes the state space of X to be $\mathbb{R}$, there is no difficulty in extending (1.1) to processes X on $\mathbb{R}^m$: then W becomes an m-dimensional Wiener process, $b(x)$ and $k(x,z)$ are m-dimensional vectors for every $x \in \mathbb{R}^m$, and $c(x)$ is an $m \times m$ matrix for every $x \in \mathbb{R}^m$; the Poisson random measure N can remain as it is. We shall call a Hunt process X on $\mathbb{R}^m$ an *Ito process* if X satisfies (1.1) for some b, c, k; (we put no conditions on b, c, k except that the integrals in (1.1) must be well-defined; we do not assume that X is the only solution of (1.1)).

§1b. *Semimartingale Hunt processes*

Here is the most interesting result of this paper. Let X be a Hunt process on $\mathbb{R}^m$, and suppose further that X is a semimartingale (that is, X can be written as the sum of a local martingale and a process of locally bounded variation -- there is no implication that either term be Markov). Then, there is an Itô process $\hat{X}$ and a positive Borel function $a \le 1$ such that

$$\hat{A}_t = \int_0^t a(\hat{X}_s)\, ds, \qquad t \ge 0, \tag{1.3}$$

is strictly increasing and continuous, and if A is the functional inverse of $\hat{A}$, then

$$X_t = \hat{X}_{A_t}, \qquad t \ge 0. \tag{1.4}$$

In other words, for every semimartingale Hunt process X on $\mathbb{R}^m$ (defined over a sufficiently large probability space), there exist deterministic functions a, b, c, k and a Wiener process W and a Poisson random measure N with the fixed mean measure $ds \cdot dz/z^2$ such that X is obtained by (1.4) from a process $\hat{X}$ that satisfies (1.1). Here, W and N are defined over the same probability space as X, and are independent of each other but not of X. If the coefficient functions b, c, k turn out to be such that $\hat{X}$ is the only solution of (1.1), then the probability law of the original process X is completely specified by the four deterministic functions a, b, c, k.

§1c. *Processes with locally bounded variation*

Let X be a Hunt process, and suppose that its every path has finite variation (that is, paths have bounded variation over every finite interval, or equivalently, every path can be written as the difference of two functions each of which is increasing in every component). Then, X is automatically a semimartingale, and the representation of §1b holds. Moreover, in this case, X satisfies (1.4) with A and $\hat{A}$ as before and the Itô process $\hat{X}$ now satisfying

$$(1.5)\qquad \hat{X}_t = \hat{X}_0 + \int_0^t \hat{b}(\hat{X}_s)\,ds + \int_0^t \int_{\mathbb{R}_0} k(\hat{X}_{s-},z)\,N(ds,dz),$$

where

$$\hat{b}(x) = b(x) - \int k(x,z)\,I_{\{|k(x,z)|\le 1\}}\,dz/z^2:$$

in other words, $c = 0$ and k is such that the third integral in (1.1) can be broken into two ordinary Stieltjes integrals.

If X is further continuous, then $k = 0$ in (1.5). But the homogeneous strong Markov processes that are solutions to (1.5) with $k = 0$ are deterministic once $\hat{X}_0$ is known (see [4] for a proof). Then, (1.3) shows that $\hat{A}$ is deterministic given $\hat{X}_0$, which implies that A is deterministic given $\hat{X}_0 = X_0$, which implies in turn that X is deterministic given X_0. In other words, there is a deterministic function $p\colon\ \mathbb{R}^m \times \mathbb{R}_+ \to \mathbb{R}^m$, which is Borel measurable in the first argument and continuous in the second, such that

$$(1.6)\qquad X_t = p(X_0,t), \qquad t \ge 0.$$

Of course, p will satisfy $p(p(x,t),u) = p(x,t+u)$. In general, p is not differentiable -- the paths of X are not necessarily differentiable even though those of $\hat{X}$ are.

§1d. *Physical interpretation*

We have just seen that any continuous strong Markov process having paths with finite variation is deterministic. For a motion to be physically plausible, the paths must be continuous and have finite variation (the latter means that the particle can travel only a finite amount during a finite time interval). Thus, there are no physically realizable strong Markov processes that are non-deterministic.

We now give a pictorial description for processes X discussed in §1c. As the recent work of TANAKA [25] indicates, the description has some merit in physics. But our main aim is to give an intuitive meaning to equations like (1.1) and (1.5).

Consider a particle moving in $\mathbb{R}^3$, let X_t denote its "velocity" at time t as shown on some "speedometer" attached to the particle. In addition to acceleration and deceleration, which change the velocity continuously, there are shocks occurring from time to time and with random magnitudes, which cause the velocity to change instantaneously.

The evolution of the velocity and shock processes is hard to describe directly. But, if time is measured on a carefully defined intrinsic time scale, then both the shocks and the velocity process become easier to characterize. We may think of the intrinsic time as the time shown on a clock attached to the particle, and whose mechanism is affected by the velocity process. Suppose the clock is such that, whenever the velocity shown on the speedometer is x, the standard time passes at the rate of $a(x)$ standard time units per unit of clock's time. Then, letting $\hat{X}_t$ denote the velocity shown on the speedometer when the clock shows t, the formula (1.3) shows that $\hat{A}_t$ is the standard time when the clock shows t. Then, A_t is what the clock shows when the standard time is t, and hence (1.4) holds.

We now describe the velocity process $(\hat{X}_t)$ as a function of the intrinsic time scale. As reckoned by the clock, the times of the shocks form some sequence (T_i) -- the T_i are not ordered. Let Z_i denote the size of the shock occurring at T_i. The pairs (T_i, Z_i) form a Poisson random measure N on $\mathbb{R} \times \mathbb{R}_0$, that is, if $N(B)$ is the number of pairs (T_i, Z_i) belonging to B, then $N(B)$ has the Poisson distribution with parameter $n(B) = \int_B ds \cdot dz/z^2$. In response to the shock (T_i, Z_i), the velocity process $\hat{X}$ jumps by the amount $k(\hat{X}_{T_i -}, Z_i)$ at the time T_i. In other words, every time a shock occurs, if the velocity just previous to the shock is x and if the magnitude of the shock is z, then the velocity jumps by the amount $k(x,z)$. Since $N(ds,dz)$ is 1 or 0 according as (s,z) equals (T_i, Z_i) for some i or not, the net change in velocity during $(0,t]$ due to shocks is

$$\sum_{T_i \le t} k(\hat{X}_{T_i -}, Z_i) = \int_0^t \int_{\mathbb{R}_0} k(\hat{X}_{s-}, z)\, N(ds,dz).$$

Finally, letting $\hat{b}(x)$ denote the "acceleration" when the velocity is x, we see that the velocity process $(\hat{X}_t)$ satisfies (1.5).

Of course, the differentiability of the continuous change and the Poissonness of the shock mechanism are both due to the use of an intrinsic time scale. Making the mean measure of N to be $ds \cdot dz/z^2$ is by choice -- it is possible to replace dz/z^2 by any other infinite σ-finite diffuse measure by changing k.

Note that there are infinitely many shocks during any time interval of positive length. However, the velocity does not have to jump every time a shock occurs: if x is a "sticky" velocity, then shocks of small enough magnitude might be unable to change it. In other words, if $\hat{X}_{t-}$ is x and a shock occurs at that time t with magnitude z, then a jump is caused only if $k(x,z) \neq 0$. In particular, if x is a holding point, then $\hat{b}(x) = 0$ and $k(x,z) = 0$ for all z in some neighborhood of the origin.

Consequently, someone observing the velocity process will not necessarily see all the shocks. Moreover, even with shocks whose times are noticed, because they cause jumps in velocity, the exact magnitudes of the schocks might not be inferrable from the velocity process: if $\hat{X}_{t-}$ is x and $\hat{X}_t$ is $y \neq x$, all that is known is that there was a shock at time t with some magnitude z satisfying $k(x,z) = y-x$, and there might be more than one such z. Because of these possibilities, the Poisson random measure N is not determined (sample pointwise) by the process $\hat{X}$. Hence, starting with $\hat{X}$, the construction of N requires us to supply the ineffective shocks ourselves as well as supplying the exact magnitudes of the shocks when they are partially known.

In mathematical terms, this requires enlarging the original probability space over which X is defined, and is the cause of much hardship -- almost all of §4 is devoted to this. A similar statement holds for more general motions, when Wiener processes are present, for the construction of dW_t when $c(X_t)$ vanishes.

§1e. *Non-semimartingale Hunt processes*

The class of semimartingale Hunt processes is very large: it contains all diffusions, all regular step processes, all the processes we have seen in applications, etc. But there are Hunt processes that are not semimartingales: if W is the Wiener process on $\mathbb{R}$, $X = |W|^{\frac{1}{2}}$ is a Hunt process that is *not* a semimartingale (see YOR [27]).

In the case of regular continuous strong Markov processes X on $\mathbb{R}$, the first step of FELLER's characterization consists of showing the existence of a strictly increasing continuous "scale" transformation $f\colon \mathbb{R} \to \mathbb{R}$ such that the Markov process $f(X)$ has the same hitting distributions as a Brownian motion. Then, $f(X)$ can be obtained from a Brownian motion by a random time change, and hence, is a martingale Markov process. FELLER's method is based on computing the probabilities of exiting an interval $[a,b]$ at a and at b starting at a point $x \in (a,b)$, and then stretching the real line in such a way that the probabilities become just as for a Brownian motion. This method does not generalize to processes on $\mathbb{R}^m$ with $m > 1$ or to processes on $\mathbb{R}$ that have jumps. But it does suggest the following program:

Given a Hunt process X, first, characterize those functions f for which $f(X)$ is a semimartingale; then, study such $f(X)$ and obtain representations for them; and finally, infer the structure of X from those of $f(X)$ for a manageable number of f's. Incidentally, in this program, the state space of X can be arbitrary.

The first problem was solved in [5]: $f(X)$ is a semimartingale if and only if f is locally the difference of two excessive functions. The second problem is solved here partially: for such f, $Y = f(X) - f(X_0)$ is a semimartingale additive functional of X, and we give characterizations for all such additive functionals that are quasi-left-continuous. The final problem, inferring the structure of X from that of $f(X)$ for sufficiently many f, largely depends on the existence of a finite collection $\{f_i\}$ of functions that separate the points in the state space of X and are such that each $f_i(X)$ is a semimartingale. This program was first advocated by KNIGHT [18], who discussed its scope under the hypothesis that there exist a sequence of excessive functions separating the points of the state space. In some regards, the present paper is the more complete application of martingale theory that was advocated by KNIGHT. It will be completed in [4].

Partly because of the considerations above, we concentrate on the representation of semimartingale additive functionals of a given process X. If X itself is a semimartingale, then $Y = X - X_0$ is a semimartingale additive functional of X and the results mentioned above for X become immediate corollaries. Moreover, one gets by-products of some interest. Suppose X is a semimartingale Markov process on $\mathbb{R}^m$ that is not time-homogeneous. Then, (T_0+t,X_t) is a time-homogeneous Markov process, and (t,X_t-X_0) is an $\mathbb{R}^{m+1}$-valued semimartingale additive functional of it. Assuming that the quasi-left-continuity etc. hold, it follows that X is obtained by (1.4), where A is the functional inverse of

$$\hat{A}_t = \int_0^t a(s,\hat{X}_s)\, ds, \qquad t \geq 0,$$

from a time-inhomogeneous Ito process $\hat{X}$ satisfying (1.1) where the coefficients now are $b(s,X_s)$, $c(s,X_s)$, and $k(s,X_{s-},z)$.

§1f. *General remarks*

Although the results we shall present appear new, the underlying concepts and techniques have been known for some time. We merely bring together various ideas from stochastic calculus, semimartingales, point processes, and Markov processes.

For stochastic calculus, our basic reference is [16]. The conversion of multivariate point processes to Poisson random measures was done by GRIGELIONIS [12], and independently, by EL KAROUI and LEPELTIER [7] and by KABANOV, LIPTSER, SHIRYAEV [17]. Representation of continuous martingales as stochastic integrals with respect to Wiener processes is a classical result. Similar representations were given for semimartingales by EL KAROUI and LEPELTIER [7]. Our only contribution, then, is in bringing these results together within the setting of a Markov process and in reconciling the differences in method. Even then, because the results appear interesting, and because the work required is too technical to be left as an exercise, we choose to present them fully in the format of a review paper.

§2. Preliminaries

In this section we give the precise conditions on the basic Markov process we consider, give some definitions, and state some preliminary results on semimartingales.

§2a. *Basic Markov process*

Throughout this paper, E is a universally measurable subset of a compact metrizable space, $\mathcal{E}$ is its Borel σ-field, and $\mathcal{E}^*$ is its σ-field of universally measurable subsets.

Let $X = (\Omega,\mathcal{F},\mathcal{F}_t,\theta_t,X_t,\mathbb{P}_x)$ be a right continuous left-hand-limited strong Markov process with infinite lifetime, with state space E, and with transition semi-group (P_t). We assume that either X is normal or θ_0 is the identity mapping on Ω. See BLUMENTHAL and GETOOR [2] for the precise meanings.

For each finite measure μ on $(E,\mathcal{E})$ we let $\mathbb{P}_\mu$ denote $\int \mu(dx)\mathbb{P}_x$. As usual, $\mathcal{F}^o_t = \sigma(X_s\colon s \le t)$, $\mathcal{F}^o = \vee_t \mathcal{F}^o_t$, $\mathcal{F}^\mu$ is the completion of $\mathcal{F}^o$ with respect to $\mathbb{P}_\mu$, $\mathcal{F}^\mu_t$ is the σ-field generated by $\mathcal{F}^o_t$ and the $\mathbb{P}_\mu$-null sets of $\mathcal{F}^\mu$, $\mathcal{F} = \cap_\mu \mathcal{F}^\mu$, and $\mathcal{F}_t = \cap_\mu \mathcal{F}^\mu_t$.

We will need to work with extensions of the process X and we need to introduce larger filtrations than $(\mathcal{F}^o_t)$. If $(\mathcal{M}^o_t)$ is such a filtration, we set $\mathcal{M}^o = \vee_t \mathcal{M}^o_t$, and we automatically assume that $\mathbb{P}_x(d\omega)$ is a transition probability from $(E,\mathcal{E}^*)$ into $(\Omega,\mathcal{M}^o)$. We construct M and (M_t) from M^o and (M^o_t) exactly the same way as F and (F_t) are obtained from F^o and (F^o_t).

(2.1) DEFINITION. We say that (M^o_{t+}) (or (M_t)) is a *strong Markov filtration* if M^o_t is a separable σ-field for every $t \geq 0$ and if for every finite stopping time T of (M^o_{t+}) the following hold:

i) $X_T \in M^o_{T+}/E$, $\theta_T \in M/M^o$;

ii) $M^o_{(T+u)+} = M^o_{T+} \vee \theta_T^{-1}(M^o_{u+})$ for all $u \geq 0$;

iii) $\mathbb{E}_\mu [\, Z\circ\theta_T \mid M^o_{T+} \,] = \mathbb{E}_{X_T} [\, Z \,]$ for all μ and all $Z \in bM^o$.

Of course (F_t) is a strong Markov filtration. If (M^o_t) is a strong Markov filtration, we have $M_{t+} = M_t$ and $(\Omega, M, M_t, \theta_t, X_t, \mathbb{P}_x)$ is a strong Markov process in the usual sense of [2], but with the additional property that (2.1iii) holds for every $Z \in bM^o$ rather than holding only for $Z \in bF^o$. If T is a stopping time of (M_t), possibly non-finite, we have

(2.2) $$\mathbb{E}_\mu [\, Z\circ\theta_T \mid M_T \,] = \mathbb{E}_{X_T} [\, Z \,] \quad \text{on } \{ T < \infty \}.$$

We are interested in obtaining the best possible measurability results in particular when $\mathbb{P}_x(d\omega)$ is a transition probability from (E,E) into (Ω, M^o) instead of being from (E,E^*) into (Ω, M^o); when $M^o = F^o$, this amounts to saying that (P_t) is a Borel semigroup. To unify the treatment for all possible cases we introduce the following.

(2.3) CONVENTION. Let X be the process described, and let (M^o_t) be a strong Markov filtration. Throughout this paper, E_0, H, H_t will satisfy one of the following three cases:

i) $\mathcal{E}_0 = \mathcal{E}^*$, $\mathcal{H}_t = \mathcal{M}_t$, $\mathcal{H} = \mathcal{M}$.

ii) $\mathcal{E}_0 = \mathcal{E}$, $\mathcal{H}_t = \mathcal{M}^o_{t+}$, $\mathcal{H} = \mathcal{M}^o$; in this case we assume that $\mathbb{P}_x(d\omega)$ is a transition probability from $(E,\mathcal{E})$ into $(\Omega,\mathcal{M}^o)$.

iii) $\mathcal{E}_0 = \mathcal{E}^e$, which is the σ-field on E generated by the α-excessive functions $(\alpha > 0)$; $\mathcal{H}_t = \mathcal{F}^e_{t+}$ where $\mathcal{F}^e_t = \sigma(f(X_s);\ s \le t,\ f \in b\mathcal{E}^e)$; $\mathcal{H} = \vee_t \mathcal{H}_t$. In this case we have $\mathcal{E} \subset \mathcal{E}_0 \subset \mathcal{E}^*$ and $\mathcal{F}^o_{t+} \subset \mathcal{H}_t \subset \mathcal{F}_t$, and we assume that X is a "right" process (see [11]).

Moreover, by an $(\mathcal{H}_t)$-adapted functional we will mean a process that is adapted to $(\mathcal{F}_t)$, $(\mathcal{F}^o_{t+})$, or $(\mathcal{F}^e_{t+})$ respectively according as the condition (i), (ii), or (iii) is in force.

Note that, with this convention, we always have

(2.4) if $f \in b\mathcal{E}_0$, then $f(X_t) \in b\mathcal{H}_t$;

(2.5) if $Z \in b\mathcal{H}$, then $x \to \mathbb{E}_x[\,Z\,]$ is $\mathcal{E}_0$-measurable.

Our basic setup consists of the strong Markov process X equipped with a strong Markov filtration $(\mathcal{H}_t)$ satisfying (2.3).

§2b. *Additive local martingales and Wiener processes*

Let P be a probability measure on $(\Omega,\mathcal{H})$. If Y is a locally square integrable right-continuous local martingale on $(\Omega,\mathcal{H},\mathcal{H}_t,P)$, we denote by $\langle Y,Y\rangle$ the predictable increasing process in the Doob-Meyer decomposition of the local submartingale Y^2. It is called the *quadratic variation* process of Y with respect to P. If Y' is another locally square integrable local martingale,

$$\langle Y,Y'\rangle = \frac{1}{4}\{\langle Y+Y',Y+Y'\rangle - \langle Y-Y',Y-Y'\rangle\}$$

is called the *quadratic covariation* process of Y and Y'. Note that, for $Y' = Y$, this reduces to the quadratic variation process for Y.

(2.6) DEFINITION. An *additive locally square integrable martingale* on (X,H_t) is a real-valued process Y that is adapted to (H_t), is right-continuous, is a locally square integrable local martingale on $(\Omega,H,H_t,\mathbb{P}_x)$ for every $x \in E$, and is additive with respect to (θ_t) (that is, for every pair (t,u), $Y_{t+u} = Y_t + Y_u \circ \theta_t$ almost surely). Moreover, we say that Y is *continuous* (resp. *quasi-left-continuous*) if it has almost surely continuous paths (resp. it is a quasi-left-continuous process under every $\mathbb{P}_x$, $x \in E$).

The content of this sub-section is essentially due to KUNITA and WATANABE [19], with the techniques of [5] enabling us to get rid of "Hypothesis (L)." For instance, Theorem (3.18) of [5] gives us the following.

(2.7) THEOREM. If Y and Y' are additive locally square integrable martingales on (X,H_t), there exists an (H_t)-adapted additive functional $\langle Y,Y'\rangle$ with finite variation that is a version of the quadratic covariation process of Y and Y' with respect to every measure $\mathbb{P}_x$. Moreover, $\langle Y,Y'\rangle$ is continuous whenever Y or Y' is quasi-left-continuous.

Our only contribution to the subject is the following theorem, where we construct the stochastic integral of the process $f(X_-)$ when f is not necessarily Borel and is only universally measurable. Readers interested only in case (ii) of Convention (2.3) may skip this, since it is then a classical result.

(2.8) THEOREM. Let Y be an additive locally square integrable martingale on (X,H_t). Let $f \in E_0$ be such that the process

$$B_t = \int_0^t f(X_{s-})^2 \, d\langle Y,Y\rangle_s, \qquad t \geq 0, \tag{2.9}$$

is almost surely finite-valued and has bounded jumps. Then, there exists an additive locally square integrable martingale N on (X,H_t) with the following property: for every $x \in E$, N is the unique (up to $\mathbb{P}_x$-indistinguishability) local martingale on $(\Omega,H,H_t,\mathbb{P}_x)$ that is a stochastic integral with respect to Y and is such that $\langle N,Y\rangle_t = \int_0^t f(X_{s-}) \, d\langle Y,Y\rangle_s$.

For the process N described in the preceding theorem, we write

$$N_t = \int_0^t f(X_{s-}) \, dY_{s-}, \qquad t \geq 0, \tag{2.10}$$

or even $\int_0^t f(X_s) \, dY_s$ when Y is continuous. Of course, when f is Borel measurable, the process $f(X_-)$ is (H_t)-predictable, the stochastic integral (2.10) is well-defined, and our theorem is a part of Theorem (3.18) of [5]. When f is not Borel measurable, the difficulty comes from the fact that $f(X_-)$ is no longer (H_t)-predictable (or even measurable), so that the results of [5] do not apply.

(2.11) REMARK. Note that such a difficulty does not arise in such expressions as (2.9): for each ω, $s \to f(X_{s-}(\omega))$ is $(R_+)^*$-measurable, and hence, $B_t(\omega)$ is well-defined. Indeed, in case (2.3i) it is a classical result that $B_t \in M_t$ (see [2] for instance); in case (2.3ii), it is evident that $B_t \in M^o_{t+}$; in case (2.3iii), it is known (see [1] for instance) that the increasing additive functional B is indistinguishable from an $(\mathcal{F}^e_{t+})$-adapted one. Hence, in all cases, there is an $(\mathcal{H}_t)$-adapted increasing additive functional such that B is $\mathbb{P}_x$-indistinguishable from it for every $x \in E$.

(2.12) REMARK. The assumption that B has bounded jumps is not necessary. The result is true without it, but the proof is more complicated. At any rate, we will need the result only when Y is quasi-left-continuous, which implies that B is continuous.

PROOF. Let $f \in \mathcal{E}^*$ and $x \in E$. Considering the measure $A \to \mathbb{E}_x[\int_0^\infty 1_A(X_{s-})\, d\langle Y,Y\rangle_s]$ on $(E,\mathcal{E})$, the universal measurability of f implies the existence of Borel measurable functions f_x and f'_x having the property that $f_x \le f \le f'_x$ and that

$$(2.13) \qquad \mathbb{E}_x\left[\int_0^\infty (f'_x - f_x)(X_{s-})\, d\langle Y,Y\rangle_s\right] = 0.$$

The process $f_x(X_-)$ is $(\mathcal{H}_t)$-predictable, and $\int_0^t f_x(X_{s-})^2\, d\langle Y,Y\rangle_s$ is $\mathbb{P}_x$-a.s. finite for every $t \ge 0$. Hence, there exists a $\mathbb{P}_x$-locally square integrable martingale N^x that is the stochastic integral $N^x_t = \int_0^t f_x(X_{s-})\, dY_s$ for the measure $\mathbb{P}_x$. Moreover, the quadratic covariances $\mathbb{P}_x$-$\langle N^x,Y\rangle$ and $\mathbb{P}_x$-$\langle N^x,N^x\rangle$ under $\mathbb{P}_x$ are

$$(2.14) \qquad \mathbb{P}_x\text{-}\langle N^x,Y\rangle_t = \int_0^t f_x(X_{s-})\, d\langle Y,Y\rangle_s,$$

$$(2.15) \qquad \mathbb{P}_x\text{-}\langle N^x,N^x\rangle_t = \int_0^t f_x(X_{s-})^2\, d\langle Y,Y\rangle_s.$$

In view of (2.13), we may replace f_x by f on right-hand-sides of (2.14) and (2.15).

Let $T_n = \inf\{\, t : B_t > n \,\}$. By Remark (2.11), we may assume that

B is $(\mathcal{H}_t)$-adapted; hence, T_n is a $(\mathcal{H}_t)$-stopping time. Since B has only bounded jumps, by (2.15) we have

$$\mathbb{E}_x[\,(N^x_{t\wedge T_n})^2\,] = \mathbb{E}_x[\,\mathbb{P}_x\text{-}\langle N^x,N^x\rangle_{t\wedge T_n}\,] = \mathbb{E}_x[\,B_{t\wedge T_n}\,] < \infty,$$

that is, $N^x_{t\wedge T_n}$ is $\mathbb{P}_x$-integrable.

Let $C \in \mathcal{M}^o$. By Lemma (3.32) of [5], there exists a bounded $(\mathcal{H}_t)$-adapted right-continuous process V that is a version of the $\mathbb{P}_x$-martingale $\mathbb{E}_x[\,1_C|\mathcal{H}_t\,]$ for every x. Moreover, by Theorem (3.12) of [5], there exists an $(\mathcal{H}_t)$-adapted process $\langle Y,V\rangle$ that is the quadratic covariation of Y and V for every measure $\mathbb{P}_x$. From (2.14) with f_x replaced by f, we have $\mathbb{P}_x\text{-}\langle N^x,V\rangle_t = \int_0^t f(X_{s-})\,d\langle Y,V\rangle_s$. Since $N^x_{t\wedge T_n}$ is $\mathbb{P}_x$-integrable, these yield

$$\mathbb{E}_x[\,N^x_{t\wedge T_n}\,1_C\,] = \mathbb{E}_x[\,N^x_{t\wedge T_n}\,V_{t\wedge T_n}\,]$$

$$= \mathbb{E}_x[\,\mathbb{P}_x\text{-}\langle N^x,V\rangle_{t\wedge T_n}\,] = \mathbb{E}_x[\,\int_0^{t\wedge T_n} f(X_{s-})\,d\langle Y,V\rangle_s\,].$$

Using again Remark (2.11) and the fact that T_n is a $(\mathcal{H}_t)$-stopping time, we obtain that $\int_0^{t\wedge T_n} f(X_{s-})\,d\langle Y,V\rangle_s = Z$ almost surely for every $\mathbb{P}_x$, where Z is a $\mathcal{H}_t$-measurable random variable. Thus, we can apply Lemmas (3.30) and (3.27) of [5] in that order to obtain right-continuous adapted processes N(n) that are $\mathbb{P}_x$-indistinguishable from $(N^x_{t\wedge T_n})_{t\geq 0}$ for every x. Putting $N_t = N(n)_t$ for $T_{n-1} < t \leq T_n$, we obtain that N is $\mathbb{P}_x$-indistinguishable from N^x for every x. Hence, N is a locally square integrable local martingale of $(X,\mathcal{H}_t)$ that has all the wanted properties, except that we do not yet know if it is additive. To show that N is additive, it suffices to reproduce the proof of (3.15vi) and (3.18vi) of [5].

The next result is a variant of the well-known orthonormalization procedure of Gram-Schmidt.

(2.16) THEOREM. Let $(Y^i)_{i\in I}$ be a collection of additive locally square integrable martingales on (X,H_t) indexed by a set I of the form $I = \{1,2,\dots,m\}$ or $I = N^* = \{1,2,\dots\}$, and assume that each Y^i is quasi-left-continuous.

a) There exists a collection $(M^i)_{i\in I}$ of additive quasi-left-continuous locally square integrable martingales on (X,H_t) and a collection $(a_{ij})_{i,j\in I}$ of E_0-measurable functions such that $a_{ij} = 0$ if $j > i$ and

$$(2.17) \qquad \langle M^i,M^j\rangle = 0 \quad \text{if } i \neq j,$$

$$(2.18) \qquad Y^i_t = \sum_{j\le i} \int_0^t a_{ij}(X_{s-})\, dM^j_s.$$

b) If A is an (H_t)-adapted increasing continuous additive functional such that $d\langle Y^i,Y^i\rangle_t \ll dA_t$ almost surely for every $i \in I$, then there exists a collection $(c_{ij})_{i,j\in I}$ of E_0-measurable functions such that $c_{ij} = 0$ if $j > i$ or if $c_{jj} = 0$ and

$$(2.19) \qquad \langle Y^i,Y^j\rangle_t = \int_0^t \Big[\sum_{k\in I} c_{ik}(X_s)c_{jk}(X_s) \Big]\, dA_s.$$

Moreover, one may choose M^i and a_{ij} in part a) such that $a_{ij} = c_{ij}$ and that, with $B_i = \{ x: c_{ii}(x) \neq 0 \}$,

$$(2.20) \qquad \langle M^i,M^i\rangle_t = \int_0^t 1_{B_i}(X_s)\, dA_s. \qquad \square$$

Of course, when we write an expression like (2.18), we mean in particular that the stochastic integrals exist in the sense of Theorem (2.8), that is, each process $\int_0^t a_{ij}(X_{s-})^2 \, d\langle M^j,M^j\rangle_s$ is finite valued for $j \le i$.

(2.21) REMARK. When I is finite, $A = \sum_i \langle Y^i,Y^i\rangle$ satisfies the hypothesis of b) above. When $I = \mathbb{N}^*$, one can again find an A satisfying the hypothesis of b): it is possible to find a sequence (N^n) of additive square integrable martingales on $(X,\mathcal{H}_t)$ that "generates" the space of all square integrable martingales and is such that $\mathbb{E}_x[\langle N^n,N^n\rangle_t] \le 1+t^2$ for every n. Then, $A' = \sum 2^{-n} \langle N^n,N^n\rangle$ is finite valued and one may take for A the continuous part of A'. See KUNITA and WATANABE [19] or MEYER [20].

(2.22) REMARK. The quasi-left-continuity of the Y^i is essential in order to have a representation like (2.18) where the integrated processes are functions of X_-.

PROOF. a) We prove the result by induction. Set $M^1 = Y^1$ and $a_{11}(x) = 1$. Suppose $M^1,\ldots,M^{n-1}$; $a_{1i},\ldots,a_{n-1,i}$ have been obtained and satisfy (2.17) and (2.18) for $i,j \le n-1$. The additive functionals $\langle Y^n,M^i\rangle$ and $\langle M^i,M^i\rangle$ are $(\mathcal{H}_t)$-adapted, and continuous, and satisfy $d\langle Y^n,M^i\rangle_t \ll d\langle M^i,M^i\rangle_t$. Hence by MOTOO's theorem (see [5], Theorem (3.55)), there exist an $\mathcal{E}_0$-measurable function a_{ni} such that $\langle Y^n,M^i\rangle_t = \int_0^t a_{ni}(X_s)\, d\langle M^i,M^i\rangle_s$, and this implies that $\int_0^t a_{ni}(X_s)^2 \cdot d\langle M^i,M^i\rangle_s$ is finite. Hence, Theorem (2.8) allows us to put

$$M_t^n = Y_t^n - \sum_{j\le n-1} \int_0^t a_{nj}(X_{s-})\, dM_s^j,$$

which defines an additive quasi-left-continuous locally square integrable martingale on (X,H_t). Using (2.17) for $i,j \le n-1$ and the definition of c_{nj} for $j \le n-1$, we obtain

$$<M^n,M^j>_t = <Y^n,M^j>_t - \int_0^t a_{nj}(X_s)\, d<M^j,M^j>_s = 0$$

for $j \le n-1$. Hence, if we set $a_{nn}(x) = 1$ and $a_{ni}(x) = 0$ for $i > n$, the collection $\{ M^1,\dots,M^n; a_{1i},\dots,a_{ni} \}$ satisfies (2.17) and (2.18) for all $i,j \le n$.

b) It follows from the construction above that $d<M^n,M^n>_t <<$ $\sum_{j\le n} d<Y^j,Y^j>_t$, which yields that $d<M^n,M^n>_t << dA_t$. Applying MOTOO's theorem once more, we obtain a collection $(b_i)_{i\in I}$ of E_0-measurable positive functions such that $<M^i,M^i>_t = \int_0^t b_i(X_s)\, dA_s$. Then, $c_{ij}(x) = a_{ij}(x)\sqrt{b_j(x)}$ satisfies (2.19) and $c_{ij} = 0$ if $j > i$ or if $c_{jj} = 0$ (this last property is because $a_{jj} \ne 1$).

Now let $\tilde{c} = (\tilde{c}_{ij})_{i,j\in I}$ be another matrix-valued E_0-measurable function that satisfies (2.19) and $\tilde{c}_{ij} = 0$ if $i < j$ or if $\tilde{c}_{jj} = 0$. We will now use matrix notation; note that all our matrices are lower-triangular so that all the products below are well-defined although I may be infinite; in this notation, (2.19) for instance reads as $<Y^i,Y^i>_t = \int_0^t \tilde{c}\tilde{c}'_{ij}(X_s)\, dA_s$, where $\tilde{c}'$ is the transpose of $\tilde{c}$. Let $b = (b_{ij})$ be defined by $b_{ij} = b_i\delta_{ij}$. We saw above that $<Y^i,Y^j>_t = \int_0^t aba'_{ij}(X_s)\, dA_s$, so we may assume that $\tilde{c}\tilde{c}' = aba'$.

Let $\tilde{B}_i = \{ x: \tilde{c}_{ii}(x) \ne 0 \}$. Let $\lambda = (\lambda_{ij})$ and $\rho = (\rho_{ij})$ be defined by $\lambda_{ij} = \delta_{ij}\, 1_{\tilde{B}_i}$ and $\rho_{ij} = \tilde{c}_{ij}$ if $c_{jj} \ne 0$ and $\rho_{ij} = \delta_{ij}$ otherwise. Since $\tilde{c}_{ij} = 0$ whenever $\tilde{c}_{jj} = 0$, we have $\tilde{c} = \rho\lambda$.

Since ρ is a lower-triangular E_0-measurable matrix with $\rho_{ii} \ne 0$, there is another E_0-measurable lower-triangular matrix d that is an inverse of ρ, i.e., $\rho d = d\rho = \delta$. Let $f = da$. We have

$$fbf' = daba'd' = d\tilde{c}\tilde{c}'d' = d\rho\lambda\lambda'\rho'd' = \lambda\lambda' = \lambda.$$

Hence, $\int_0^t f_{ij}(X_s)^2 \, d\langle M^j,M^j\rangle_s \le \int_0^t \lambda_{ij}(X_s)\, dA_s$ is finite, and we can set

$$\tilde{M}_t^i = \sum_{j\le i} \int_0^t f_{ij}(X_{s-})\, dM_s^j.$$

Since $fbf' = \lambda$, we have $\langle \tilde{M}^i,\tilde{M}^j\rangle_t = \int_0^t \lambda_{ij}(X_s)\, dA_s$, that is, $(\tilde{M}^i)_{i\in I}$ satisfies (2.17) and (2.20) with $\tilde{B}_i$ replacing B_i. Finally, $\tilde{c}\lambda\tilde{c}' = \tilde{c}\tilde{c}'$ satisfies (2.19) by hypothesis, so we may put

$$\tilde{Y}_t^i = \sum_{j\le i} \int_0^t \tilde{c}_{ij}(X_{s-})\, d\tilde{M}_s^j.$$

Since $\tilde{c}f = \tilde{c}da = a$ and (2.18) holds, we have $\tilde{Y}^i = Y^i$ and the theorem is proved.

In the next definition, the index set I is either of the form $I = \{ 1,\dots,m \}$ or $I = \mathbb{N}^* = \{ 1,2,\dots \}$.

(2.23) DEFINITION. A *Wiener process* $W = (W^i)_{i\in I}$ over $(X,\mathcal{H}_t)$ is a collection of $(\mathcal{H}_t)$-adapted continuous additive processes W^i such that $W_0^i = 0$ and the increments $W_{t+u}^i - W_t^i$, $i \in I$, are independent of each other and of $\mathcal{H}_t$ and have means 0 and variances u under every $\mathbb{P}_x$, $x \in E$.

The following characterization may be found in [16] and does not require that $(\mathcal{H}_t)$ be a strong Markov filtration.

(2.24) PROPOSITION. A process $W = (W^i)_{i\in I}$ is a Wiener process over $(X,\mathcal{H}_t)$ if and only if every W^i is an additive continuous local martingale on $(X,\mathcal{H}_t)$ and $\langle W^i,W^j\rangle_t = \delta_{ij}\, t$ for all $i,j \in I$ and $t \ge 0$.

§2c. *Additive random measures and Poisson random measures*

We collect here some useful facts about random measures. For a complete treatment see [16], and also [5] for random measures on Markov processes.

Let $(D,\mathcal{D})$ be a Lusin space with its σ-field of Borel sets. A *random measure* Γ on $\mathbb{R}_+ \times D$ is a positive transition kernel $\Gamma(\omega;dt,dy)$ from $(\Omega,\mathcal{H})$ into $(\mathbb{R}_+ \times D, \mathcal{R}_+ \otimes \mathcal{D})$. It is said to be *integer valued* if

(2.25) i) $\Gamma(\omega;C) \in \{0,1,\dots,+\infty\}$ for all $\omega \in \Omega$ and $C \in \mathcal{R}_+ \otimes \mathcal{D}$,

ii) $\Gamma(\omega;\{0\}\times D) = 0$, $\Gamma(\omega;\{t\}\times D) \le 1$ for all $\omega \in \Omega$, $t \ge 0$.

Given a random measure Γ and a set $B \in \mathcal{D}$, we let Γ^B denote the increasing process $(t,\omega) \to \Gamma^B_t(\omega) = \Gamma(\omega;[0,t]\times B)$. The random measure Γ is said to be $(\mathcal{H}_t)$-optional (resp. predictable) if Γ^B is $(\mathcal{H}_t)$-optional (resp. predictable) for every $B \in \mathcal{D}$.

Let P be a probability measure on $(\Omega,\mathcal{H})$. Let Γ be an $(\mathcal{H}_t)$-optional random measure for which there exists a $\mathcal{D}$-measurable partition (B_n) of D such that every process Γ^{B_n} is locally integrable with respect to P. Then, there exists an $(\mathcal{H}_t)$-predictable random measure G, unique up to a P-null set, called the *dual predictable projection* of Γ, such that the following two equivalent conditions are satisfied:

(2.26) $\Gamma^B - G^B$ is a local martingale on $(\Omega,\mathcal{H},\mathcal{H}_t,P)$ (that is, G^B is the dual predictable projection of Γ^B) for every $B \in \mathcal{D}$ such that Γ^B is locally integrable.

$$(2.27) \qquad E[\int \Gamma(dt,dy)\, U(t,y)\,] = E[\int G(dt,dy)\, U(t,y)\,]$$

for all $\mathcal{P} \otimes \mathcal{D}$-measurable positive functions U on $\Omega \times \mathbb{R}_+ \times D$, where $\mathcal{P}$ is the $(\mathcal{H}_t)$-predictable σ-algebra on $\Omega \times \mathbb{R}_+$.

(2.28) DEFINITION. An *additive random measure* on $(X, \mathcal{H}_t)$ is an $(\mathcal{H}_t)$-optional random measure Γ such that

i) for every $B \in \mathcal{D}$, Γ^B is an additive process (possibly with infinite values),

ii) there exists a $\mathcal{D}$-measurable partition (B_n) of D for which Γ^{B_n} is $\mathbb{P}_x$-locally integrable for every $x \in E$.

(2.29) EXAMPLE. Let (Y_t) be an $(\mathcal{H}_t)$-adapted process taking values in $D = \mathbb{R}^m$ or $D = \mathbb{R}^{\mathbb{N}}$, and suppose that its paths are right continuous and left-hand-limited. Let ΔY denote the process defined by $\Delta Y_0 = 0$, $\Delta Y_t = Y_t - Y_{t-}$ for $t > 0$. Then,

$$\Gamma(\omega; dt, dy) = \sum_{s>0} I_{\{\Delta Y_s(\omega) \neq 0\}} \varepsilon_{(s, \Delta Y_s(\omega))}(dt, dy)$$

(where ε_a denotes the Dirac measure putting its unit mass at a) is an additive integer valued random measure on $\mathbb{R}_+ \times D$.

(2.30) THEOREM. a) If Γ is an additive random measure on $(X, \mathcal{H}_t)$, then there exists an $(\mathcal{H}_t)$-predictable additive random measure G on $(X, \mathcal{H}_t)$ that is a version of the dual predictable projection of Γ under every measure $\mathbb{P}_x$. Moreover, if H is a positive $(\mathcal{H}_t)$-predictable process and if g is a positive function on $E \times D$ that is $\mathcal{E}^* \otimes \mathcal{D}$-measurable (or even $(\mathcal{E} \otimes \mathcal{D})^*$-measurable), then for every $x \in E$ we have

$$(2.31) \quad \mathbb{E}_x[\int \Gamma(dt, dy) H_t g(X_{t-}, y)] = \mathbb{E}_x[\int G(dt, dy) H_t g(X_{t-}, y)]$$

b) Suppose further that Γ is integer valued and quasi-left-continuous (that is, $\mathbb{P}_x\{\ \omega: \Gamma(\omega;\{T(\omega)\}\times D) > 0\ \} = 0$ for every finite predictable stopping time T of (H_t) and every x). Then, there exists an (H_t)-adapted continuous increasing additive functional A and a positive kernel $K(x,dy)$ from (E,E_0) into $(D,\mathcal{D})$ with the following properties:

(2.32) there exists an $E_0 \otimes \mathcal{D}$-measurable strictly positive function f such that $\int K(x,dy)f(x,y) < \infty$ for every $x \in E$;

(2.33) the dual predictable projection G of Γ is given by

$$G(\omega;dt,dt) = dA_t(\omega)\ K(X_t(\omega),dy).$$

Moreover, if A' is another (H_t)-adapted continuous increasing functional such that $dA_t \ll dA'_t$, then there exists another kernel K' satisfying (2.32) such that (2.33) holds with A' and K' replacing A and K.

PROOF. All the statements are proved in [5], Theorems (6.6) and (6.19), except for the last assertion in (a) and for the existence of f in (2.32).

Let $g \in (E\otimes\mathcal{D})^*$ and $x \in E$. There exist two $E \otimes \mathcal{D}$-measurable functions g_x and g'_x such that $g_x \leq g \leq g'_x$ and

(2.34) $$\mathbb{E}_x [\ \int [\ \Gamma(dt,dy) + G(dt,dy)\ ]\ H_t\ (g'_x-g_x)(X_{t-}-y)\ ] = 0.$$

Both the function $H_t(\omega)\cdot g_x(X_{t-}(\omega),y)$ and $H_t(\omega)\cdot g'_x(X_{t-}(\omega),y)$ are $P \otimes \mathcal{D}$-measurable in (ω,t,y), and thus, (2.31) is satisfied with g_x

and g'_x because of (2.27) and the definition of G. Hence, (2.34) implies that g also satisfies (2.31).

To show the existence of f in (2.32), let (A,K) be a pair satisfying (2.33), and let (B_n) be the partition described in (2.18ii). Then, $F = \bigcap_n \{ x: K(x,B_n) < \infty \}$ is $\mathcal{E}_0$-measurable and satisfies $\int_0^t 1_F(X_s)\, dA_s = A_t$ almost surely. Thus, by replacing $K(x,\cdot)$ by $1_F(x)K(x,\cdot)$ we may assume that $F = E$ without altering (2.33). Then, (2.32) is fulfilled with $f(x,y) = \sum_n 2^{-n} K(x,B_n)^{-1} \cdot 1_{B_n}(y)$ where $0^{-1} = 1$.

We turn now to the construction of some stochastic integrals with a result in the same vein as Theorem (2.8).

(2.35) THEOREM. Let Γ be an integer valued and quasi-left-continuous additive random measure on $(X,\mathcal{H}_t)$ and let G be its dual predictable projection as constructed in (2.30). Let $g \in \mathcal{E}_0 \otimes \mathcal{D}$ be such that the process

$$B_t = \int_0^t \int_D G(ds,dy)\, [\, g(X_{s-},y)^2 \wedge |g(X_{s-},y)|\,], \qquad t \geq 0,$$

is almost surely finite-valued. Then, there exists an additive local martingale N on $(X,\mathcal{H}_t)$ with the following property: for every $x \in E$, N is the unique (up to $\mathbb{P}_x$-indistinguishability) local martingale on $(\Omega,\mathcal{H},\mathcal{H}_t,\mathbb{P}_x)$ that is a compensated sum of jumps and is such that ΔN is $\mathbb{P}_x$-indistinguishable from

$$\tilde{g}_t(\omega) = \int_D \Gamma(\omega;\{t\}\times dy)\, g(X_{t-}(\omega),y). \tag{2.36}$$

The process N above is denoted by

(2.37) $$N_t = \int_0^t \int_D g(X_{s-},y) \, [\, \Gamma(ds,dy) - G(ds,dy)\,],$$

and coincides with the same ordinary integral whenever $|g(X_{t-},y)|$ is integrable with respect to both treasures Γ and G.

PROOF. If g were $\mathcal{E} \otimes \mathcal{D}$-measurable, then $g(X_{t-}(\omega),y)$ would be $\mathcal{P}(\mathcal{H}_t) \otimes \mathcal{D}$-measurable in (ω,t,y) and the result would follow from Proposition (6.13) of [5].

Let $g \in \mathcal{E}_0 \otimes \mathcal{D} \subset (\mathcal{E}\otimes\mathcal{D})^*$. For each $x \in E$, there exists two $\mathcal{E} \times \mathcal{D}$-measurable functions g_x and g'_x satisfying $g_x \le g \le g'_x$ and (2.34). Then,

$$\int_0^t \int_D G(ds,dy) \, [\, g_x(X_{s-},y)^2 \wedge |g_x(X_{s-},y)| \,] = B_t$$

$\mathbb{P}_x$-almost surely. Since $B_t<\infty$, $\mathbb{P}_x$-a.s., it follows that $g_x(X_{t-},y)$ is $\mathbb{P}_x$-integrable with respect to $\Gamma - G$ (see [16]), and we denote by N^x the corresponding stochastic integral: N^x is $(\mathcal{H}_t)$-adapted and is the only $\mathbb{P}_x$-local martingale that is a compensated sum of jumps having ΔN^x given by (2.36) with g there replaced by g_x. Moreover, (2.34) implies that ΔN^x and $\tilde{g}$ are $\mathbb{P}_x$-indistinguishable, and hence,

$$A(n)_t = \sum_{s\le t} \tilde{g}_s \, 1_{\{\, |\tilde{g}_s| \ge 1/n \,\}}$$

is $\mathbb{P}_x$-indistinguishable from $\sum_{s\le t} \Delta N_s^x \, 1_{\{\, |\Delta N_s^x| \ge 1/n \,\}}$, which has $\mathbb{P}_x$-locally integrable variation. Clearly, $A(n)$ is additive; hence, the same argument as in Remark (2.11) shows that $A(n)$ is indistinguishable from an $(\mathcal{H}_t)$-adapted additive functional, which we again denote by $A(n)$. Since $A(n)$ has $\mathbb{P}_x$-locally integrable variation for every x, by the results of [5], it admits a dual predictable projection $\tilde{A}(n)$ that is (H_t)-adapted, additive, and continuous (since $A(n)$ is quasi-left-continuous). Thus, $N(n) = A(n) - \tilde{A}(n)$ is an additive local martingale on (X,H_t). Moreover, by the definitions of $N(n)$ and N^x, the sequence $(N(n)_t)_{n\geq 0}$ converges in $\mathbb{P}_x$-measure toward N_t^x for every $t \geq 0$. Hence, the arguments in the proof of Proposition (6.13) of [5] show that there exists an additive (H_t)-adapted right-continuous process N that is $\mathbb{P}_x$-indistinguishable from N^x for every $x \in E$. This process N satisfies all the requirements of the theorem.

Let ν be a positive σ-finite measure on $(D,\mathcal{D})$ and let $n(dt,dy) = dt\,\nu(dy)$.

(2.38) DEFINITION. A *Poisson random measure* over (X,H_t) *with mean measure* n is an integer valued additive random measure N over (X,H_t) such that, for every integer k and disjoint sets $A_1,\dots,A_k$ in $\mathcal{R}_+ \otimes \mathcal{D}$ and contained in $(t,\infty) \times D$, the random variables $N(A_1),\dots,N(A_k)$ are independent of each other and of H_t and have means $n(A_1),\dots,n(A_k)$ under every measure $\mathbb{P}_x$, $x \in E$.

If N is a Poisson random measure over (X,H_t), then for every $A \in \mathcal{R}_+ \otimes \mathcal{D}$ with $n(A) < \infty$ the random variable $N(A)$ has the Poisson distribution with mean $n(A)$ under every $\mathbb{P}_x$. The following characterization may be found in [16] and does not require (H_t) to be strong Markov.

(2.39) PROPOSITION. a) An integer valued additive random measure N over (X,H_t) is a Poisson random measure with mean measure $n(dt,dy) = dt\,\nu(dy)$ if and only if its dual predictable projection is the (deterministic) measure n.

b) If N is a Poisson random measure and W is a Wiener process both over (X,H_t), then N and W are independent under every $\mathbb{P}_x$.

§2d. *Additive semimartingales*

Let P be a probability measure on (Ω,H). An m-dimensional semimartingale on (Ω,H,H_t,P) is a process $Y = (Y^i)_{i\le m}$ whose every component Y^i is the sum of a local martingale and a process with bounded variation over every finite interval. We define

$$(2.40) \qquad Y_t^e = \sum_{s\le t} \Delta Y_s \, I_{\{|\Delta Y_s| > 1\}}, \quad t \ge 0,$$

where ΔY is defined as in (2.19). This is the sum of jumps of size exceeding 1 in magnitude. Then, Y^e is a right continuous pure jump process with finitely many jumps in any finite interval. The semimartingale $Y - Y^e$ has bounded jumps, and hence, has a unique decomposition $Y - Y^e = Y_0 + Y^b + M$ where Y^b is a predictable process with *bounded* variation on every finite interval, M is a local martingale, and $Y_0^b = M_0 = 0$. Moreover, M can be decomposed as $M = Y^c + Y^d$ where Y^c is a *continuous* local martingale, Y^d is a "purely *discontinuous*" local martingale (a compensated sum of jumps), and $Y_0^c = Y_0^d = 0$. We end up with

$$(2.41) \qquad Y = Y_0 + Y^b + Y^c + Y^d + Y^e;$$

and this decomposition is unique up to a P-null set. All processes in (2.41) are m-dimensional, their i^{th} components are denoted by Y^{ib}, Y^{ic}, etc.

Let $B = Y^b$ in the decomposition (2.41), let $C = (C^{ij})_{i,j\le m}$

where $C^{ij} = \langle Y^{ic}, Y^{jc} \rangle$, and let G be the dual predictable projection of the random measure Γ defined by (2.19), all for the semimartingale Y. Then, (B,C,G) is called the triplet of *local characteristics* of Y. This triplet is defined uniquely up to a P-null set, and in many cases, it specifies the probability law of Y.

(2.42) DEFINITION. An *additive semimartingale* over (X,H_t) is an m-dimensional process Y that is adapted to (H_t), is additive with respect to (θ_t), and is a semimartingale with respect to every $\mathbb{P}_x$, $x \in E$.

The following is proved in [5], Theorems (3.18) and (6.24).

(2.43) THEOREM. Let Y be an m-dimensional additive semimartingale over (X,H_t).

a) There exists a decomposition $Y = Y^b + Y^c + Y^d + Y^e$ that is the decomposition (2.41) for Y with respect to every measure $\mathbb{P}_x$, and each one of Y^b, Y^c, Y^d, Y^e is (H_t)-adapted and additive (note that $Y_0 = 0$ by additivity).

b) There is a triplet (B,C,G) that is the triplet of local characteristics of Y with respect to every $\mathbb{P}_x$; B and C are (H_t)-predictable additive processes, G is an (H_t)-predictable additive random measure.

The next theorem is obtained from Theorems (2.16) and (2.30) here and Theorem (6.25) in [5].

2.44) THEOREM. Let Y be an m-dimensional additive semimartingale over (X,H_t), and suppose that it is quasi-left-continuous (that is,

$Y_{T_n} \to Y_T$ almost surely for every increasing sequence (T_n) of stopping times with finite limit T). Then, there exist

(i) an $(\mathcal{H}_t)$-adapted continuous increasing additive functional A,

(ii) an $\mathcal{E}_0$-measurable $\mathbb{R}^m$-valued function $b = (b_i)_{i \le m}$,

(iii) an $\mathcal{E}_0$-measurable $m \times m$ lower triangular matrix-valued function $c = (c_{ij})_{i,j \le m}$ with $c_{ij} = 0$ whenever $c_{jj} = 0$,

(iv) and a positive kernel $K(x,dy)$ from $(E,\mathcal{E}_0)$ into $(\mathbb{R}^m, \mathcal{R}^m)$ having $K(x,\{0\}) = 0$ for all $x \in E$ and for which (2.32) is satisfied with $D = \mathbb{R}^m$, such that

(2.45)
$$B_t = \int_0^t b(X_s)\, dA_s, \qquad C_t = \int_0^t cc'(X_s)\, dA_s,$$
$$G(dt,dy) = dA_t\, K(X_t, dy)$$

define a version (B,C,G) of the triplet of local characteristics of Y under every $\mathbb{P}_x$, $x \in E$.

Moreover, if $\tilde{A}$ is another process like A in (i) with $dA_t \ll d\tilde{A}_t$, then there exist $\tilde{b}$, $\tilde{c}$, $\tilde{K}$ satisfying (ii) - (iv) such that (2.45) holds with $\tilde{A}$, $\tilde{b}$, $\tilde{c}$, $\tilde{K}$ replacing A, b, c, K.

We will call (A,b,c,K) a *system of local characteristics* for Y. Note that we may always replace A_t by $t + A_t$. Thus, we may and do assume that A is strictly increasing and $dt \ll dA_t$.

§2e. *Markov extensions*

As was mentioned in the introduction, we seek representations for semimartingale additive functionals in terms of Wiener processes and Poisson random measures. It is possible that, for some x, the original space $(\Omega, \mathcal{H}, \mathbb{P}_x)$ is not large enough to hold the auxiliary processes required. For this reason, we need to enlarge the spaces

$(\Omega, H, \mathbb{P}_x)$ in such a way that the original process X remains Markov. We now make this notion of enlargement precise.

Let (Ω', H', P') be an auxiliary probability space, set

$$(2.46) \qquad (\tilde{\Omega}, \tilde{H}, \tilde{\mathbb{P}}_x) = (\Omega, H, \mathbb{P}_x) \times (\Omega', H', P'),$$

and let π be the projection mapping $(\omega, \omega') \to \omega$ from $\tilde{\Omega}$ to Ω. For any random variable Z defined on Ω (e.g. $Z = X_t$), we denote by the same letter Z its natural extension $Z \circ \pi$ to $\tilde{\Omega}$: $Z(\omega, \omega') = Z(\omega)$, and similarly for Z defined on Ω'.

Let $(\tilde{H}_t)$ be a filtration on $(\tilde{\Omega}, \tilde{H})$ such that $\tilde{H}_t = \tilde{H}_{t+}$, and let $(\tilde{\theta}_t)$ be a semi-group of transformations on $\tilde{\Omega}$. Set $\tilde{X} = (\tilde{\Omega}, \tilde{H}, \tilde{H}_t, \tilde{\theta}_t, X_t, \tilde{\mathbb{P}}_x)$.

(2.47) DEFINITION. The pair $(\tilde{X}, \tilde{H}_t)$ is called a *strong Markov extension* of (X, H_t) if

i) $\pi \in \tilde{H}_t / H_t$ and $\pi \circ \tilde{\theta}_t = \theta_t \circ \pi$ for all $t \geq 0$,

ii) for every $Z \in b\tilde{H}$ and every finite stopping time T of $(\tilde{H}_t)$, $Z \circ \tilde{\theta}_T$ is measurable with respect to the completion of $\tilde{H}$ and

$$(2.48) \qquad \tilde{\mathbb{E}}_x [\, Z \circ \tilde{\theta}_T \mid \tilde{H}_T \,] = \tilde{\mathbb{E}}_{X_T} [\, Z \,].$$

It is evident that, then $\tilde{X}$ is a strong Markov process and has the same transition function as X. The filtration $(\tilde{H}_t)$ would be a strong Markov filtration of X except that it does not necessarily satisfy (2.1ii). We omit the (very easy) proof of the next proposition.

(2.49) PROPOSITION. A strong Markov extension of a strong Markov extension of (X, H_t) is again a strong Markov extension of (X, H_t).

(2.50) PROPOSITION. Let $(\tilde{X},\tilde{H}_t)$ be a strong Markov extension of (X,H_t).

a) Let $Y = (Y^i)_{i\in I}$ be a collection of continuous additive local martingales over (X,H_t) with quadratic covariations $<Y^i,Y^j>$. Then, the processes Y^i are still continuous additive local martingales over $(\tilde{X},\tilde{H}_t)$ with the same quadratic covariations $<Y^i,Y^j>$.

b) Let Γ be an additive random measure over (X,H_t) with dual predictable projection G. Then, Γ is still an additive random measure on $(\tilde{X},\tilde{H}_t)$ with the same dual predictable projection G.

c) Let Y be an additive semimartingale over (X,H_t) with local characteristics triplet (B,C,G) and let $Y = Y^b + Y^c + Y^d + Y^e$ be its decomposition (2.41). Then, Y is an additive semimartingale over $(\tilde{X},\tilde{H}_t)$ with the same triplet (B,C,G) of local characteristics and the same decomposition $Y = Y^b + Y^c + Y^d + Y^e$.

PROOF. By (2.47i), any (H_t)-adapted (resp. predictable) process or random measure that is additive with respect to (θ_t) on Ω is also, when considered over $\tilde{\Omega}$, $\tilde{H}_t$-adapted (resp. predictable) and additive with respect to $(\tilde{\theta}_t)$. It is therefore sufficient (see [16], §IX-2-c) to prove that any bounded martingale on $(\Omega,H,H_t,\mathbb{P}_x)$ is also a martingale on $(\tilde{\Omega},\tilde{H},\tilde{H}_t,\tilde{\mathbb{P}}_x)$. For this it is enough to show that $\tilde{\mathbb{E}}_x[\, I_A \mid \tilde{H}_t\,] = \mathbb{E}_x[\, I_A \mid H_t\,]$ for every $t \geq 0$ and $A \in H$. By (2.1ii), it is sufficient to prove this equality for A of the form $A = A' \cap \theta_t^{-1}(A'')$ with $A' \in H_t$ and $A'' \in H$. Now (2.1iii) and (2.47ii) yield $\mathbb{E}_x[\, I_A \mid H_t\,] = I_{A'}\, \mathbb{P}_{X_t}(A'')$ and $\tilde{\mathbb{E}}_x[\, I_A \mid \tilde{H}_t\,] = I_{A'}\, \tilde{\mathbb{P}}_{X_t}(A'')$, and $\tilde{\mathbb{P}}_{X_t}(A'') = \mathbb{P}_{X_t}(A'')$ by (2.47i) and (2.46).

(2.51) REMARK. If $\tilde{Y}$ is a continuous additive local martingale over $(\tilde{X},\tilde{H}_t)$, one may define its quadratic variation $<\tilde{Y},\tilde{Y}>$ to be independent of $\tilde{\mathbb{P}}_x$. But one does not know if $\langle\tilde{Y},\tilde{Y}\rangle$ is additive, because $(\tilde{H}_t)$ is not necessarily a strong Markov filtration of $\tilde{X}$. However, in c) above, we start with an additive Y on (X,H_t), which implies that $\langle Y,Y\rangle$ is additive on (X,H_t) and therefore on $(\tilde{X},\tilde{H}_t)$. These remarks apply to statements b) and c) above as well.

Using the characterizations (2.24) and (2.39), we obtain the following.

(2.52) COROLLARY. A Wiener process (resp. A Poisson random measure) over (X,H_t) remains a Wiener process (resp. a Poisson random measure) over every strong Markov extension of (X,H_t). (Note that this does not require (H_t) to be a strong Markov filtration of X.)

(2.53) REMARK. It is possible to introduce a more general notion of strong Markov extension by dropping the product form (2.46) and assuming only that $\tilde{\Omega} = \Omega \times \Omega'$ and $\pi \in \tilde{H}/H$ and $\mathbb{P}_x = \tilde{\mathbb{P}}_x \pi^{-1}$. However, all the extensions we will construct are of the pleasant product form of (2.46). On the other hand, we will not be able to make the extended filtrations $(\tilde{H}_t)$ to have such pleasant product forms as $\tilde{H}_t = \bigcap_{s>t} (H_s \otimes H'_s)$.

§3. Representation Theorems

Throughout this section, X is a Markov process with a strong Markov filtration (H_t) as described in §2a. Our objective is to list all our results in a precise manner and to give all the easy proofs (leaving the complicated ones to §4).

§ 3a. *The fundamental result*

Let I be a finite or countable index set, which can always be written as $I = \{1,\dots,m\}$ or $I = \mathbb{N}^* = \{1,2,\dots\}$. Let $(D,\mathcal{D})$ be a Lusin space. Let Δ be an extra point added to D to form $D_\Delta = D \cup \{\Delta\}$, and let $\mathcal{D}_\Delta$ be the σ-field on D_Δ generated by $\mathcal{D}$. Finally, let ν be some fixed positive σ-finite infinite measure on $\mathbb{R}$ without any atoms. Consider the following objects.

(3.1) Let $Y = (Y^i)_{i\in I}$ be a collection of continuous additive local martingales on $(X,\mathcal{H}_t)$ such that $d\langle Y^i,Y^i\rangle_t \ll dt$ almost surely for all $i \in I$. Let $c = (c_{ij})_{i,j\in I}$ be the collection of $\mathcal{E}_0$-measurable functions whose existence and properties are given by Theorem (2.16b) with $A_t = t$.

(3.2) Let Γ be an additive integer-valued random measure on $\mathbb{R}_+ \times D$ defined over $(X,\mathcal{H}_t)$. Let G be its dual predictable projection, and assume that $dG_t^B \ll dt$ almost surely for every $B \in \mathcal{D}$ such that $t \to G_t^B = G([0,t]\times B)$ is locally integrable. By Theorem (2.30) this assumption is equivalent to the existence of a positive kernel K from $(E,\mathcal{E}_0)$ into $(D,\mathcal{D})$ such that (2.32) holds, and

$$(3.3) \qquad G(dt,dy) = dt\ K(X_t,dy) \quad \text{a.s.}$$

The following lemma is almost classical (see e.g. EL KAROUI and LEPELTIER [7], or Lemma (4.15) in JACOD [16], another proof will be provided in §4 later).

(3.4) LEMMA. Let K be a positive kernel from $(E,\mathcal{E}_0)$ into $(D,\mathcal{D})$ such that (2.32) is satisfied. Then, there exists a measurable function $k: (E \times R, \mathcal{E}_0 \otimes \mathcal{R}) \to (D_\Delta, \mathcal{D}_\Delta)$ such that, for every $x \in E$ and $B \in \mathcal{D}$,

$$K(x,B) = \int_{\mathbb{R}} \nu(dz)\, 1_B \circ k(x,z). \tag{3.5}$$

(3.6) REMARK. The function k depends on the choice of the fixed measure ν on $\mathbb{R}$. If $K(x,D) < \infty$ for every x, taking $\nu(dz) = dz$ gives the simplest representation for k. If $D = \mathbb{R}\setminus\{0\}$ and $K(x,\mathbb{R}\setminus[-\varepsilon,\varepsilon]) < \infty$ for every $x \in E$ and $\varepsilon > 0$, then taking $\nu(dz) = dz/z^2$ for $z \neq 0$ is the most convenient computationally (e.g. when K is associated with a one-dimensional semimartingale by (2.45)). Then, with $1/0 = \infty$ and $\Delta = 0$, we have

$$k(x,z) = \begin{cases} \inf\{\, y>0:\ K(x,(y,\infty))^{-1} > z \,\} & \text{if } z \geq 0, \\ \sup\{\, y<0:\ K(x,(-\infty,y])^{-1} < z \,\} & \text{if } z < 0. \end{cases}$$

However, any ν will do as long as it is σ-finite infinite and without atoms. In fact, instead of working with a measure ν on $\mathbb{R}$, we could work with any such measure ν' on any Lusin space D'.

The following is the major result.

(3.7) THEOREM. Consider the objects described in (3.1) and (3.2), and let k satisfy (3.5). Then, there exists a strong Markov extension $(\tilde{X},\tilde{\mathcal{H}}_t)$ of $(X,\mathcal{H}_t)$ supporting a Wiener process $\tilde{W} = (\tilde{W}^i)_{i\in I}$ and a Poisson random measure $\tilde{N}$ on $\mathbb{R}_+ \times \mathbb{R}$ with mean measure $n(dt,dz) = dt\,\nu(dz)$ such that

$$(3.8) \qquad Y_t^i = \sum_{j\in I} \int_0^t c_{ij}(X_s)\, d\tilde{W}_s^j, \qquad i \in I,$$

$$(3.9) \qquad \Gamma(B) = \int \tilde{N}(ds,dz)\, 1_B(s,k(X_{s-},z)), \qquad B \in \mathcal{R}_+ \otimes \mathcal{D},$$

almost surely under each $\tilde{\mathbb{P}}_x$. Moreover, we can take $(\tilde{X},\tilde{\mathcal{H}}_t) = (X,\mathcal{H}_t)$ provided that $c_{ii}(x)$ be non-zero for all $x \in E$ and $i \in E$ and the measure $K(x,\cdot)$ in (3.3) be infinite and without atoms for all $x \in E$.

We postpone the proof to the next section. In (3.8), we have a finite sum because $c_{ij} = 0$ if $j > i$, and the stochastic integral is the one defined in Theorem (2.8) except that this is over the process $(\tilde{X},\tilde{\mathcal{H}}_t)$.

Expression (3.9) may be re-stated as follows: for almost every (ω,ω') in $\tilde{\Omega}$ (with respect to $\tilde{\mathbb{P}}_x$ for every x), the measure $\Gamma(\omega;ds,dy)$ is the image of the measure $\tilde{N}((\omega,\omega');ds,dz)$ under the mapping $(s,z) \to (s,k(X_{s-}(\omega),z))$.

(3.10) PROPOSITION. Let $(\tilde{X},\tilde{\mathcal{H}}_t)$ be a strong Markov extension of $(X,\mathcal{H}_t)$ supporting a Wiener process $\tilde{W} = (\tilde{W}^i)_{i\in I}$ and a Poisson random measure $\tilde{N}$ on $\mathbb{R}_+ \times \mathbb{R}$ with mean measure $ds\,\nu(dz)$.

a) Let $c = (c_{ij})_{i,j\in I}$ be a collection of $\mathcal{E}_0$-measurable functions such that $c_{ij} = 0$ for all $i < j$ and that $\int_0^t c_{ij}(X_s)^2 ds < \infty$ $\mathbb{P}_x$-almost surely for each $x \in E$. Then, the right-hand-side of (3.8) defines a process $\tilde{Y}^i$ which is the same under every $\tilde{\mathbb{P}}_x$ and which is a continuous additive local martingale of $(\tilde{X},\tilde{\mathcal{H}}_t)$ with $\langle\tilde{Y}^i,\tilde{Y}^j\rangle_t = \int_0^t cc'(X_s)\,ds$ (where c' is the transpose of c).

b) Let k be a measurable function from $(E \times \mathbb{R},\ \mathcal{E}_0 \otimes \mathcal{R})$ into $(D_\Delta,\mathcal{D}_\Delta)$ such that, if $\tilde{\Gamma}$ is defined by the right-hand-side of (3.9), there exists a $\mathcal{D}$-measurable partition (B_n) of D with $\tilde{\Gamma}([0,t]\times B_n)<\infty$

almost surely for every n and $t \geq 0$. Then, $\tilde{\Gamma}$ is an additive integer-valued random measure over $(\tilde{X},\tilde{H}_t)$ whose dual predictable projection is $\tilde{G}(dt,dy) = dt\ K(X_t,dy)$ where K is defined by (3.5).

PROOF. The statement a) is just a re-statement of Theorem (2.8). To prove b), let $B \in \mathcal{D}$ be contained in one of the B_n's. Then, $\tilde{\Gamma}^B$ is additive, and the argument in Remark (2.11) shows that $\tilde{\Gamma}^B$ is indistinguishable from an $(\tilde{H}_t)$-adapted additive functional of $\tilde{X}$. Hence, $\tilde{\Gamma}$ is an additive integer-valued random measure over $(\tilde{X},\tilde{H}_t)$. If H is a positive $(\tilde{H}_t)$-predictable process, then (2.30a) applied to $\tilde{N}$ and $g(x,z) = 1_B(x,k(x,z))$ shows that

$$\tilde{\mathbb{E}}_x\left[\int_0^t H_s\, d\tilde{\Gamma}_s^B\right] = \tilde{\mathbb{E}}_x\left[\int_0^t H_s\, ds \int_{\mathbb{R}} \nu(dz)\, 1_B(X_{s-},k(X_{s-},z))\right]$$

$$= \tilde{\mathbb{E}}_x\left[\int_0^t H_s\, d\tilde{G}_s^B\right]$$

by the definition of $\tilde{G}$ (and by the fact that X_{t-} and X_t differ for at most countably many t, and therefore, $\tilde{G}(dt,dy) = dt\ K(X_{t-},dy)$). Hence, $\tilde{G}^B$ is finite-valued and is indistinguishable from an $(\tilde{H}_t)$-predictable process by Remark (2.11), and it follows that $\tilde{G}$ is the dual predictable projection of $\tilde{\Gamma}$.

The following corollary provides a converse to Theorem (3.7).

(3.11) COROLLARY. Under the hypotheses of Proposition (3.10), further assume that $\tilde{Y}^i(\omega,\omega')$ and $\tilde{\Gamma}((\omega,\omega');\cdot)$ are free of ω', that is, $\tilde{Y}^i(\omega,\omega') = Y^i(\omega)$ and $\tilde{\Gamma}((\omega,\omega');\cdot) = \Gamma(\omega;\cdot)$ for some Y^i and Γ. Suppose that $H_t = M_t$, that is, suppose (2.3i) holds. Then, $(Y^i)_{i\in I}$ and Γ satisfy the assumptions of (3.1) and (3.2).

PROOF. a) Any bounded martingale on $(\Omega, \mathcal{H}, \mathcal{H}_t, \mathbb{P}_\mu)$ is also a martingale on $(\tilde{\Omega}, \tilde{\mathcal{H}}, \tilde{\mathcal{H}}_t, \tilde{\mathbb{P}}_\mu)$. This was proved in the course of proving (2.50) for $\mu = \varepsilon_x$, and extends to arbitrary μ.

b) Let $Z \in b\mathcal{H}$ be such that it is $\tilde{\mathcal{H}}_t$-measurable when considered as a variable on $\tilde{\Omega}$. Then, a) above implies that, for every $V \in b\mathcal{H}$,

$$\mathbb{E}_\mu[\, Z\, V\,] = \tilde{\mathbb{E}}_\mu[Z\, V] = \tilde{\mathbb{E}}_\mu[Z\, \tilde{\mathbb{E}}_\mu[\, V \mid \tilde{\mathcal{H}}_t\,]] = \mathbb{E}_\mu[Z\, \mathbb{E}_\mu[V \mid \mathcal{H}_t\,]].$$

Hence, $Z \in b\mathcal{H}_t^\mu$, and since $\mathcal{H}_t = \bigcap_\mu \mathcal{H}_t^\mu$, we have $Z \in b\mathcal{H}_t$.

c) It follows from b) that Y^i and Γ are adapted to $(\mathcal{H}_t)$. They are clearly additive (use (2.47i)), and Γ is an integer-valued random measure. Since Y^i is a continuous local martingale on $(\tilde{\Omega}, \tilde{\mathcal{H}}, \tilde{\mathcal{H}}_t, \tilde{\mathbb{P}}_x)$ and is $(\mathcal{H}_t)$-adapted, it is a local martingale on $(\Omega, \mathcal{H}, \mathcal{H}_t, \mathbb{P}_x)$. Now the conclusion follows from a "change of space" technique (see e.g. [16]), since $\langle \tilde{Y}^i, \tilde{Y}^j \rangle$ and $\tilde{G}$ are actually defined on Ω and are $(\mathcal{H}_t)$-predictable.

(3.12) REMARK. In the preceding corollary, the assumption that (2.3i) holds was made to ensure that Y and Γ are adapted to $(\mathcal{H}_t)$. Under (2.3ii) or (2.3iii) one can prove that Y and Γ are indistinguishable from Y' and Γ' that are $(\mathcal{H}_t)$-adapted, but the proof is rather long. Thus, the preceding corollary holds under any one of the conventions of (2.3).

§3b. *Additive semimartingales*

We now specialize the results of the preceding sub-section. Let $Y = (Y^i)_{i \le m}$ be an m-dimensional additive semimartingale over $(X, \mathcal{H}_t)$, suppose it is quasi-left-continuous, and let (A,b,c,K) be a system of local characteristics for it as defined in Theorem (2.44). The measure ν on $\mathbb{R}$ is as described in §3a, and k is as defined in Lemma (3.4) where we take $D = \mathbb{R}_0^m = \mathbb{R}^m \setminus \{0\}$ and $\Delta = 0$. The following is the main result; the condition on A will be removed later, in Theorem (3.26).

(3.13) THEOREM. Suppose $A_t = t$ identically. Then, there exists a strong Markov extension $(\tilde{X}, \tilde{H}_t)$ of (X, H_t) supporting a Wiener process $\tilde{W} = (\tilde{W}^i)_{i \le m}$ and a Poisson random measure $\tilde{N}$ on $\mathbb{R}_+ \times \mathbb{R}$ with mean measure $ds\, \nu(dz)$ such that

$$(3.14) \quad Y_t = \int_0^t b(X_s)\, ds + \int_0^t c(X_s)\, d\tilde{W}_s$$

$$+ \int_0^t \int_{\mathbb{R}} k(X_{s-}, z)\, I_{\{|k(X_{s-},z)| \le 1\}}\, [\tilde{N}(ds,dz) - ds\, \nu(dz)]$$

$$+ \int_0^t \int_{\mathbb{R}} k(X_{s-}, z)\, I_{\{|k(X_{s-},z)| > 1\}}\, \tilde{N}(ds,dz)$$

almost surely under every $\tilde{\mathbb{P}}_x$. The decomposition (2.41) of Y as $Y = Y^b + Y^c + Y^d + Y^e$ does not depend on x, is the same on (X, H_t) and $(\tilde{X}, \tilde{H}_t)$, and is given by the right-hand-side of (3.14) in that order. Moreover, we can take $(\tilde{X}, \tilde{H}_t) = (X, H_t)$ provided that the rank of the matrix $c(x)$ be m for all x and that the measure $K(x, \cdot)$ be infinite and without atoms for all x.

REMARK. On the right hand side of (3.14), the first and the last terms are ordinary integrals and are defined pathwise for each $\tilde{\omega} \in \tilde{\Omega}$. The second and third terms are multi-dimensional stochastic integrals as constructed in (2.8) and (2.35). We emphasize that the last three terms are independent of ω' even though they are defined on $\tilde{\Omega} = \Omega \times \Omega'$.

PROOF. Let $Y = Y^b + Y^c + Y^d + Y^e$ be the decomposition (2.41) of Y on (X,H_t). By Proposition (2.50), this decomposition remains the same on $(\tilde{X},\tilde{H}_t)$, and by (2.43), is the same under every $\tilde{\mathbb{P}}_x$.

The additive continuous local martingale $Y^c = (Y^{ic})_{i\le m}$ satisfies (3.1). The additive integer-valued random measure Γ associated with Y through (2.29) with $D = \mathbb{R}_0^m$ satisfies (3.2). Hence, applying Theorem (3.7), we obtain a strong Markov extension $(\tilde{X},\tilde{H}_t)$ supporting a Wiener process $\tilde{W} = (\tilde{W}^i)_{i\le m}$ and a Poisson random measure $\tilde{N}$ such that (3.9) and (3.8) hold with Y^{ic} replacing Y^i.

Thus, $Y_t^b = \int_0^t b(X_s)\,ds$ by the definition of Y^b, and $Y_t^c = \int_0^t c(X_s)\,d\tilde{W}_s$ as we have just shown. By (2.40) and the definition of Γ we have

$$Y_t^e = \int_0^t \int_{\mathbb{R}_0^m} y\, I_{\{|y|>1\}}\, \Gamma(ds,dy);$$

and hence, from the representation (3.9) for Γ, it follows that Y^e is equal to the last term on the right-hand-side of (3.14). Finally, Y^d is a purely discontinuous local martingale on $(\tilde{\Omega},\tilde{H},\tilde{H}_t,\tilde{\mathbb{P}}_x)$ whose jumps are given by

$$\Delta Y_t^d = \int_{\mathbb{R}_0^m} y\, I_{\{|y|\le 1\}}\, \Gamma(\{t\},dy),$$

which by (3.9) is equal to

$$\int_{\mathbb{R}} k(X_{t-},z)\, I_{\{|k(X_{t-},z)|\le 1\}}\, \tilde{N}(\{t\},dz).$$

Hence, by Theorem (2.5), the third term on the right-hand-side of (3.14) is well-defined and is almost surely equal to Y^d.

The remaining assertion regarding the possibility of taking

$(\tilde{X},\tilde{H}_t) = (X,H_t)$ is immediate from Theorem (3.7).

(3.15) PROPOSITION. Let $(\tilde{X},\tilde{H}_t)$ be a strong Markov extension of (X,H_t) that supports a Wiener process $\tilde{W} = (\tilde{W}^i)_{i\le m}$ and a Poisson random measure $\tilde{N}$ with mean measure $dt\, \nu(dz)$ on $\mathbb{R}_+ \times \mathbb{R}$. Let the functions b, c, k be such that the right-hand-side of (3.14) makes sense under every $\tilde{\mathbb{P}}_x$. Then, the right-hand-side of (3.14) defines a process $\tilde{Y}$ on $\tilde{\Omega}$ which is the same under every $\tilde{\mathbb{P}}_x$ and is an additive quasi-left-continuous semimartingale over $(\tilde{X},\tilde{H}_t)$. Its decomposition (2.41) as $\tilde{Y}^b + \tilde{Y}^c + \tilde{Y}^d + \tilde{Y}^e$ is the same for all measures $\tilde{\mathbb{P}}_x$ and is given by the successive terms on the right-hand-side of (3.14). Its local characteristics $\tilde{B}$, $\tilde{C}$, $\tilde{G}$ are given by (2.45) with $A_t = t$, that is,

$$\text{(3.16)} \qquad \tilde{B}_t = \int_0^t b(X_s)\, ds, \qquad \tilde{C}_t = \int_0^t cc'(X_s) ds,$$

$$\tilde{G}(B) = \int 1_B \circ k(X_t,z)\, dt\, \nu(dz).$$

PROOF. Let $\tilde{Y}^b$, $\tilde{Y}^c$, $\tilde{Y}^d$, $\tilde{Y}^e$ be the four successive terms on the right-hand-side of (3.14) respectively. Since $\tilde{Y}^b$ and $\tilde{Y}^e$ are ordinary integrals, their definitions do not depend on $\tilde{\mathbb{P}}_x$, they have bounded variation over every finite interval, they are additive by the homogeneity of X and the additivity of $\tilde{N}$, and the argument in Remark (2.11) shows that one may consider them to be $(\tilde{H}_t)$-adapted. By (3.10a), $\tilde{Y}^c$ is an additive continuous local martingale on $(\tilde{X},\tilde{H}_t)$. Since $\tilde{N}$ is additive and since its dual predictable projection is $dt\, \nu(dz)$, Theorem (2.35) applies to show that $\tilde{Y}^d$ is an additive purely discontinuous local martingale on $(\tilde{X},\tilde{H}_t)$. Hence, $\tilde{Y}$ is an additive semimartingale over $(\tilde{X},\tilde{H}_t)$. Since $\Delta\tilde{Y}^b = \Delta\tilde{Y}^c = 0$ and

$\Delta\tilde{Y}^d \le 1$, $\tilde{Y}^e$ and $\tilde{Y}$ satisfy (2.40), and it follows that $\tilde{Y}^b + \tilde{Y}^c + \tilde{Y}^d + \tilde{Y}^e$ is the decomposition (2.41) for $\tilde{Y}$.

Let $\tilde{B}$, $\tilde{C}$, $\tilde{G}$ be the local characteristics of $\tilde{Y}$, and let $\tilde{\Gamma}$ be the measure associated with $\tilde{Y}$ through (2.29). By construction , $\tilde{B} = \tilde{Y}^b$ is given by (3.16). By (3.10a), $\tilde{C}$ is given by (3.16). Finally, a simple computation shows that $\tilde{\Gamma}$ is equal to the right-hand-side of (3.9), and hence, (3.10b) implies that $\tilde{G}$ is given by (3.16). In particular, $\tilde{G}(\{t\}\times\mathbb{R}_0^m) = 0$ for every t, which implies that $\tilde{Y}$ is quasi-left-continuous.

As a corollary, we obtain a converse to Theorem (3.13), showing in particular that $A_t = t$ is a necessary assumption in order to obtain a representation such as (3.14). The proof is similar to that of (3.11).

(3.17) COROLLARY. Under the hypotheses of Proposition (3.15), further assume that $\tilde{Y}(\omega,\omega')$ is free of ω', that is, assume that $\tilde{Y}(\omega,\omega') = Y(\omega)$ for some Y. Also assume that $H_t = M_t$ (i.e. (2.3i) holds). Then, Y is an additive quasi-left-continuous semimartingale on $(\tilde{X},\tilde{H}_t)$ whose local characteristics are given by (2.45) with $A_t = t$.

(3.18) REMARK. Let $(\tilde{X},\tilde{H}_t)$ be a strong Markov extension supporting a Wiener process $\tilde{W}$ and a Poisson random measure $\tilde{N}$ with mean measure $dt\ \nu(dz)$. Let $h\colon \mathbb{R} \to \mathbb{R}^m$ be a Borel function satisfying $\int \nu(dz)(|h(z)|^2 \wedge 1) < \infty$. Then,

$$\tilde{Z}_t = \tilde{W}_t + \int_0^t \int_{\mathbb{R}} h(z)\, I_{\{|h(z)|\le 1\}}\, [\tilde{N}(ds,dz)-ds\ \nu(dz)]$$

$$+ \int_0^t \int_{\mathbb{R}} h(z)\, I_{\{|h(z)|>1\}}\, \tilde{N}(ds,dz)$$

defines an additive m-dimensional semimartingale $\tilde{Z}$ over $(\tilde{X},\tilde{H}_t)$. It is also a process with stationary and independent increments, with drift 0, diffusion matrix equal to the identity matrix, and Lévy measure νh^{-1}, all this under every $\tilde{\mathbb{P}}_x$. But $\tilde{Z}$ is not independent of X; instead, $(X,\tilde{Z}) = (\tilde{\Omega},\tilde{H},\tilde{H}_t,\tilde{\theta}_t,X_t,\tilde{Z}_t,\tilde{\mathbb{P}}_x)$ is a Markov additive process in the sense of ÇINLAR [3]. It may also be viewed as a semi-direct Markov product in the sense of JACOD [15] where either component X or $\tilde{Z}$ can be taken to be the "first component".

Proposition (3.15) and Corollary (3.17) explain the nature of the right-hand-side of (3.14). We go back to the main result, Theorem (3.13), and re-state it in the particular case where Y has finite variation (that is, Y has bounded variation over every finite interval). In this case, the stochastic integrals can be dispensed with. The setup is that of Theorem (3.13).

(3.19) PROPOSITION. Suppose $A_t = t$ identically, and suppose that Y has finite variation. Then, there exists a strong Markov extension $(\tilde{X},\tilde{H}_t)$ of (X,H_t) supporting a Poisson random measure $\tilde{N}$ on $\mathbb{R}_+ \times \mathbb{R}$ with mean measure $dt\,\nu(dz)$ such that

$$Y_t = \int_0^t \hat{b}(X_s)ds + \int_0^t \int_{\mathbb{R}} k(X_{s-},z)\,\tilde{N}(ds,dz) \tag{3.20}$$

almost surely under every $\tilde{\mathbb{P}}_x$, where

$$\hat{b}(x) = b(x) - \int_{\mathbb{R}} \nu(dz)\,k(x,z)\,I_{\{|k(x,z)|\le 1\}}. \tag{3.21}$$

PROOF. Since Y has finite variation, we have $c = 0$, and the second term on the right-hand-side of (3.14) vanishes. For the same reason, the third term has finite variation; and since the measures

$\tilde{N}(\omega;ds,dz)$ and $ds\,\nu(dz)$ are singular, this third term can be split into two ordinary integrals, one with respect to $\tilde{N}(ds,dz)$ and the other with respect to $ds\,\nu(dz)$. Thus, the proof follows from Theorem (3.13) through a rearrangement.

Finally, we remove the condition that $A_t = t$ from the main result, Theorem (3.13). The setup is that preceding (3.13). In addition, we let $\hat{A}$ be the functional inverse of A, and let $(\hat{X},\hat{H}_t)$ and $\hat{Y}$ be the processes obtained from X and Y respectively by the random time change using A as the clock:

$$\hat{A}_u = \inf\{\ t: A_t > u\ \}, \tag{3.22}$$

$$\hat{X}_u = X_{\hat{A}_u},\quad \hat{F}_u = F_{\hat{A}_u},\quad \hat{\theta}_u = \theta_{\hat{A}_u},\quad \hat{H}_u = H_{\hat{A}_u},\quad \hat{Y}_u = Y_{\hat{A}_u}. \tag{3.23}$$

Since A is a strictly increasing continuous additive functional of X with $\lim_{t\to\infty} A_t = \infty$, $\hat{X} = (\Omega,F,\hat{F}_u,\hat{\theta}_u,\hat{X}_u,\mathbb{P}_x)$ is a Markov process with the same properties as X, and $(\hat{H}_u)$ is a strong Markov filtration for $\hat{X}$. Thus, $(\hat{X},\hat{H}_u)$ satisfy the conditions listed for (X,H_t) in §2a. Further, $(\hat{Y}_u)$ is an additive quasi-left-continuous semimartingale over $(\hat{X},\hat{H}_t)$: recall that semimartingaleness property is preserved under time changes, quasi-left-continuity is preserved since A is strictly increasing and continuous, and additivity is immediate.

Let $(\hat{B},\hat{C},\hat{G})$ be the triplet of local characteristics for $\hat{Y}$ over $(\hat{X},\hat{H}_t)$. We have, from (2.45),

$$\hat{B}_u = B_{\hat{A}_u} = \int_0^{\hat{A}_u} b(X_s)\, dA_s = \int_0^u b(\hat{X}_s)\, ds, \tag{3.24}$$

and similarly,

(3.25) $$\hat{C}_u = \int_0^u cc'(\hat{X}_s)\, ds, \qquad \hat{G}(dt,dy) = du\, K(\hat{X}_u,dy).$$

Thus, the process $\hat{Y}$ over $(\hat{X},\hat{\mathcal{H}}_t)$ admits $(t;b,c,K)$ as a system of local characteristics, and Theorem (3.13) applies. Now the following should be obvious.

(3.26) THEOREM. a) There is a strong Markov extension $(\tilde{X},\tilde{\mathcal{H}}_t)$ of $(\hat{X},\hat{\mathcal{H}}_t)$ supporting a Wiener process $\tilde{W}$ on $\mathbb{R}^m$ and a Poisson random measure $\tilde{N}$ on $\mathbb{R}_+ \times \mathbb{R}$ with mean measure $ds\, \nu(dz)$ such that

(3.27) $$\hat{Y}_t = \int_0^t b(\hat{X}_s)ds + \int_0^t c(\hat{X}_s)\, d\tilde{W}_s$$

$$+ \int_0^t \int_{\mathbb{R}} k(\hat{X}_{s-},z)\, I_{\{|k(\hat{X}_{s-},z)|\leq 1\}}\, [\tilde{N}(ds,dz)-ds\, \nu(dz)]$$

$$+ \int_0^t \int_{\mathbb{R}} k(\hat{X}_{s-},z)\, I_{\{|k(X_{s-},z)|>1\}}\, \tilde{N}(ds,dz)$$

$\tilde{\mathbb{P}}_x$-almost surely for every x.

b) Y is obtained from $\hat{Y}$ by

(3.28) $$Y_t = \hat{Y}_{A_t}.$$

(3.29) REMARK. It is clear that the decomposition (2.41) of Y as $Y = Y^b + Y^c + Y^d + Y^e$ is the same for all $\mathbb{P}_x$ and is given by $Y^b_t = \hat{Y}^b_{A_t}, \ldots, Y^e_t = \hat{Y}^e_{A_t}$ where $\hat{Y}^b, \ldots, \hat{Y}^e$ are the successive terms on the right-hand-side of (3.27). Also, the extension involved can be dispensed with if the rank of $c(x)$ is m for all x and if $K(x,\cdot)$ is infinite and without atoms, just as before in (3.13).

(3.30) REMARK. In the special case where Y has finite variation, $\hat{Y}$ has finite variation and (3.19) applies. Thus, in that case, the preceding theorem holds with (3.27) replaced by

$$\hat{Y}_t = \int_0^t \hat{b}(\hat{X}_s)ds + \int_0^t \int_{\mathbb{R}} \tilde{N}(ds,dz)\, k(\hat{X}_{s-},z), \tag{3.31}$$

where $\hat{b}$ is as given by (3.21).

§3c. *Markov processes*

We now describe the implications of the results above for the Markov process X itself supposing that the state space is $E = \mathbb{R}^m$. Here $X = (\Omega, F, F_t, \theta_t, X_t, \mathbb{P}_x)$ is as described in §2a. The process X is said to be a *semimartingale Hunt process* if it is quasi-left-continuous and is a semimartingale with respect to $\mathbb{P}_x$ for every x. Then, $(Y_t) = (X_t - X_0)$ is a quasi-left-continuous semimartingale additive functional of X, and therefore, the results of the preceding subsection apply. Let (A,b,c,K) be a system of local characteristics for $Y = X - X_0$. Finally, as we had remarked earlier (see the last paragraph of §2d) we may, and do, assume that $dt \ll dA_t$ almost surely; which implies that

$$t = \int_0^t a(X_s)\, dA_s \tag{3.32}$$

for some positive E_0-measurable function a by MOTOO's Theorem ([5], (3.55)).

An *Itô process* is a semimartingale Hunt process that admits $(A;b,c,K)$ with $A_t = t$ as a system of local characteristics. This definition was introduced in [5]. The following justifies the definition of Itô processes by showing that they are indeed the processes introduced by ITO [14] under some extra conditions on b, c, K to ensure existence and uniqueness for the solution process. Note that we have already used this characterization theorem in §1 in connection with the usage "Itô process" there.

(3.33) THEOREM. Let the process X have state space $E = \mathbb{R}^m$, and set $M_t^\circ = F_t^\circ$. Then, X is an Itô process if and only if there exists a strong Markov extension $(\tilde{X}, \tilde{H}_t)$ of (X, H_t) supporting a Wiener process $\tilde{W}$ on $\mathbb{R}^m$ and a Poisson random measure $\tilde{N}$ on $\mathbb{R}_+ \times \mathbb{R}$ with mean measure $dt\,\nu(dz)$ such that (X_t) satisfies

$$(3.34)\qquad X_t = X_0 + \int_0^t b(X_s)ds + \int_0^t c(X_s)\,d\tilde{W}_s$$

$$+ \int_0^t \int_{\mathbb{R}} k(X_{s-},z)\, I_{\{|k(X_{s-},z)|\le 1\}}[\tilde{N}(ds,dz) - ds\,\nu(dz)]$$

$$+ \int_0^t \int_{\mathbb{R}} k(X_{s-},z)\, I_{\{|k(X_{s-},z)|>1\}}\, \tilde{N}(ds,dz)$$

$\tilde{\mathbb{P}}_x$-almost surely for every $x \in E$ for some E_0-measurable functions b, c, k.

PROOF is immediate from Theorem (3.13), Proposition (3.15), and Corollary (3.17) applied to $(Y_t) = (X_t - X_0)$.

Every diffusion process is an Itô process with $k = 0$; every process with stationary and independent increments is an Itô process with $b(x) = b_0$, $c(x) = c_0$, $k(x,z) = k_0(z)$ independent of x; every regular step process is an Itô process with $c = 0$ and $\hat{b} = 0$ where $\hat{b}$ is as in (3.21). On the other hand, if $X_t = |W_t|$ where W is a Brownian motion on $\mathbb{R}$, then X is a semimartingale Hunt process but is not an Itô process. The following is the characterization for semimartingale Hunt processes.

(3.35) THEOREM. Let X have state space $E = \mathbb{R}^m$. Then, X is a semimartingale Hunt process if and only if there exists a strictly increasing continuous (H_t)-adapted additive functional A such that, putting

$$\hat{A}_u = \inf\{ t: A_t > u \}, \tag{3.36}$$

$$\hat{X}_u = X_{\hat{A}_u}, \quad \hat{H}_u = H_{\hat{A}_u}, \quad \hat{\theta}_u = \theta_{\hat{A}_u} \tag{3.37}$$

yields an Itô process $\hat{X} = (\Omega, H, \hat{H}_u, \hat{\theta}_u, \hat{X}_u, \mathbb{P}_x)$. Moreover, then, X is obtained from the Itô process $\hat{X}$ through

$$X_t = \hat{X}_{A_t}, \quad H_t = \hat{H}_{A_t}, \quad \theta_t = \hat{\theta}_{A_t}, \tag{3.38}$$

and we may further assume that, for some E_0-measurable function a with $0 \le a \le 1$,

$$\hat{A}_u = \int_0^u a(\hat{X}_s)\, ds, \quad u \ge 0. \tag{3.39}$$

(3.40) REMARK. In other words, every semimartingale Hunt process is obtained from an Itô process by a random time change, and the time change can be assumed to be of a simple form (that is, through a strictly increasing continuous additive functional $\hat{A}$ of the Itô process with the simple form (3.39) above.

(3.41) REMARK. If X is a Hunt process whose paths are of finite variation, then it is automatically a semimartingale and the preceding theorem applies. Since the time change leading to an Itô process is strictly increasing, that Itô process has again paths of finite variation. Thus, the Itô process $\hat{X}$ will satisfy

$$(3.42)\quad \hat{X}_t = \hat{X}_0 + \int_0^t \hat{b}(\hat{X}_s)ds + \int_0^t \int_{\mathbb{R}} \tilde{N}(ds,dz)\, k(\hat{X}_{s-},z)$$

where $\tilde{N}$ is a Poisson random measure on $\mathbb{R}_+ \times \mathbb{R}$ over a strong Markov extension $(\tilde{X},\tilde{H}_t)$ of $(\hat{X},\hat{H}_t)$, where $\hat{b}$ is obtained from b and k through (3.21). In the further special case where the Hunt process X has paths that are continuous and of finite variation, we must have $k = 0$, and $(\hat{X}_t)$ and therefore (X_t) are deterministic; see [4].

PROOF of (3.35). a) Suppose X is a semimartingale Hunt process, and let $(A;b,c,K)$ be a system of local characteristics for it. We assume as we may, that A is strictly increasing and continuous and (H_t)-adapted. Let $\hat{X} = (\Omega,H,\hat{H}_u,\hat{\theta}_u,\hat{X}_u,\mathbb{P}_x)$ be obtained by (3.36) and (3.37). Since the semimartingale property is invariant under time changes, $(\hat{X}_t)$ is a semimartingale over $(\Omega,H,\hat{H}_u,\mathbb{P}_x)$ for every x. It is classical that $\hat{X}$ is again a Hunt process since A is strictly increasing and continuous. Thus, $\hat{X}$ is a semimartingale Hunt process. Let $(\hat{B},\hat{C},\hat{G})$ be the triplet of its local characteristics; then (3.24) and (3.25) hold and show that $\hat{X}$ admits $(t;b,c,K)$ as a system of local characteristics, and by definition, is an Itô process.

b) Since A is strictly increasing and continuous, $\tilde{A}$ is a strictly increasing continuous additive functional of $\hat{X}$, and $A_t = \inf\{ u: \hat{A}_u > t \}$. Thus, the time change (3.37) is invertible, and is obtained from $\hat{X}$ by (3.38).

c) If $\hat{X}$ is an Itô process and X is related to $\hat{X}$ through (3.36) and (3.37) for some strictly increasing continuous additive functional A of X, then (3.38) holds and shows that X must be a semimartingale Hunt process (by exactly the same arguments as in a) above).

d) Finally, since we are assuming that $dt \ll dA_t$, there exists an E_0-measurable function a with $0 \le a \le 1$ such that (3.32) holds. Thus, putting $t = \hat{A}_u$ in (3.32),

$$\hat{A}_u = \int_0^{\hat{A}_u} a(X_s)\, dA_s = \int_0^u a(\hat{X}_s)\, ds.$$

(3.43) REMARK. If the coefficients b, c, k are such that the equation (3.34) has a unique solution, then that solution process is an Itô process, and its probability law is determined by the three deterministic functions b, c, k. Then, the probability law of any semimartingale Hunt process obtained from it is completely specified by the four deterministic functions a, b, c, k where a defines the clock through (3.39). Unfortunately, there is a large gap between the known necessary conditions on b, c, k (mainly measurability) and the known sufficient conditions on b, c, k to ensure uniqueness of solutions to (3.34) (mainly Lipschitz continuity). We refer to STROOCK [19] and to STROOCK and VARADHAN [21] for the latter matters of sufficiency.

Let X be a semimartingale Hunt process admitting (A;b,c,K) as a system of local characteristics, and let $\hat{X}$ be the Itô process obtained from X by the random time change using A as the clock (that is, by (3.36) and (3.39)). Let $\tilde{N}$ be the Poisson random measure on $\mathbb{R}_+ \times \mathbb{R}$ with mean measure du $\nu(dz)$ defined over the strong Markov extension $(\tilde{X},\tilde{H}_t)$ of $(\hat{X},\hat{H}_t)$. Then, for $\tilde{\mathbb{P}}_x$-almost every $\tilde{\omega} = (\omega,\omega') \in \tilde{\Omega}$,

$$\text{(3.44)} \qquad \Delta X_t(\omega) = \int k(X_{t-}(\omega),z)\, \tilde{N}((\omega,\omega');\{A_t(\omega)\}\times dz).$$

Thus, for every positive Borel function f on $E \times E$,

$$\text{(3.45)} \quad \sum_{s\le t} f(X_{s-},X_s)\, I_{\{X_{s-}\ne X_s\}} = \int_0^{A_t} \int_{\mathbb{R}} \tilde{N}(ds,dz)\, f(\hat{X}_{s-},\hat{X}_{s-}+k(\hat{X}_{s-},z))$$

almost surely ($\tilde{\mathbb{P}}_x$, for every x). Taking expectations on both sides of (3.45), and noting the time change formulas,

$$\text{(3.46)} \qquad E_x[\sum_{s\le t} f(X_{s-},X_s)\, I_{\{X_{s-}\ne X_s\}}]$$

$$= E_x[\int_0^t dA_s \int_{\mathbb{R}} \nu(dz)\, f(X_s,X_s+k(X_s,z)).$$

This shows that X admits (A,L) as a Lévy system where

$$\text{(3.47)} \qquad L(x,B) = \int_{\mathbb{R}} \nu(dz)\, 1_B(x+k(x,z)).$$

In fact, (3.45) and therefore (3.46) and (3.47) hold for any Hunt process with state space $E = \mathbb{R}^m$, whether it is a semimartingale or not. The following states the most general result on the jump structure of Hunt processes.

(3.48) THEOREM. Let X be a Hunt process with state space E, and let (A,L) be a Lévy system for X such that A is strictly increasing and continuous and $\lim_{t\to\infty} A_t = +\infty$ (such a system exists always).

a) The process $\hat{X} = (\Omega,F,\hat{F}_u,\hat{\theta}_u,\hat{X}_u,\mathbb{P}_x)$ defined by (3.36) and

and (3.37) is a Hunt process which admits (t,L) as a Lévy system, and X is related to $\hat{X}$ by (3.38).

b) Suppose E is a Lusin space. Then there is a measurable function l: $(E \times \mathbb{R}, \mathcal{E}_0 \otimes \mathcal{R}) \to (E_\Delta, \mathcal{E}_\Delta)$ such that

$$(3.49) \qquad L(x,B) = \int \nu(dz)\, 1_B \circ l(x,z) \qquad x \in E, \quad B \in \mathcal{E}.$$

c) Suppose E is a Lusin space and let $\hat{\Gamma}$ be the integer-valued additive random measure on $\mathbb{R}_+ \times E \times E$ defined by

$$(3.50) \quad \hat{\Gamma}(dt,dx,dy) = \sum_{s>0} I_{\{\hat{X}_{s-} \neq \hat{X}_s\}}\, \varepsilon_{(s,\hat{X}_{s-},\hat{X}_s)}(dt,dx,dy).$$

Then, there exists a strong Markov extension $(\tilde{X}, \tilde{H}_t)$ of $(\hat{X}, \hat{H}_t)$ supporting a Poisson random measure $\tilde{N}$ on $\mathbb{R}_+ \times \mathbb{R}$ with mean measure dt $\nu(dz)$ such that

$$(3.51) \quad \hat{\Gamma}(B) = \int \tilde{N}(ds,dz)\, 1_B(s, X_{s-}, l(X_{s-},z)), \qquad B \in \mathcal{R}_+ \otimes \mathcal{E} \otimes \mathcal{E},$$

almost surely ($\tilde{\mathbb{P}}_x$, for every x).

d) Suppose $E = \mathbb{R}^m$. Then, there is a measurable function k: $(E \times \mathbb{R}, \mathcal{E}_0 \otimes \mathcal{R}) \to (E, \mathcal{E})$ such that (3.47) holds; and there is a strong Markov extension $(\tilde{X}, \tilde{H}_t)$ of $(\hat{X}, \hat{H}_t)$ supporting a Poisson random measure $\tilde{N}$ on $\mathbb{R}_+ \times \mathbb{R}$ with mean measure ds $\nu(dz)$ such that

$$(3.52) \qquad \sum_{s>0} 1_B(s, X_s - X_{s-}) = \int \tilde{N}(ds,dz)\, 1_B \circ k(X_{s-},z)$$

for all $B \in \mathcal{R}_+ \otimes \mathcal{E}$ with $0 \notin B$, almost surely ($\mathbb{P}_x$, for every x).

PROOF. Statement a) is proved exactly as in Theorem (3.35). Next, let $\hat{\Gamma}$ be defined by (3.50), and let $\hat{G}$ be its dual predictable projection. Then, for any $\hat{\mathcal{P}} \otimes \mathcal{E} \otimes \mathcal{E}$-measurable function U (where $\hat{\mathcal{P}}$ is the $(\hat{\mathcal{H}}_t)$-predictable σ-algebra on $\Omega \times \mathbb{R}_+$), we have

$$\mathbb{E}_x[\int \hat{G}(ds,dx,dy)U(s,x,y)] = \mathbb{E}_x[\int \hat{\Gamma}(ds,dx,dy)U(s,x,y)]$$

$$= \mathbb{E}_x[\sum_{s>0} U(s,\hat{X}_{s-},\hat{X}_s)\, I_{\{\hat{X}_{s-} \neq \hat{X}_s\}}]$$

$$= \mathbb{E}_x[\int ds \int L(\hat{X}_s,dy)\, U(s,\hat{X}_s,y)]$$

$$= \mathbb{E}_x[\int ds \int \hat{K}(\hat{X}_s;dx,dy)\, U(s,x,y)]$$

with

(3.53) $$\hat{K}(x';dx,dy) = \varepsilon_{x'}(dx)\, L(x',dy).$$

Let E be a Lusin space. Then, Lemma (3.4) applies to the kernel $\hat{K}$ defined by (3.53) to yield the existence of a $\mathcal{E}_0 \otimes \mathcal{R}$-measurable function $\hat{k}$: $E \times \mathbb{R} \to E_\Delta \times E_\Delta$ such that (3.5) holds for $\hat{K}$ and $\hat{k}$. Noting the form (3.53) of $\hat{K}$, we see that

(3.54) $$\hat{k}(x,z) = (x, l(x,z))$$

for some l: $E \times \mathbb{R} \to E_\Delta$ satisfying (3.49) as claimed in b).

Statement c) is now immediate from Theorem (3.7) applied to the process $(\hat{X},\hat{\mathcal{H}}_t)$ and the random measure $\hat{\Gamma}$ whose dual predictable projection is $dt\, \hat{K}(\hat{X}_t;dx,dy)$, and therefore in (3.9) k needs to be replaced by $\hat{k}$ above in (3.54).

When $E = \mathbb{R}^m$, if we let Γ be the image of $\hat{\Gamma}$ under the mapping

$(t,x,y) \to (t,y-x)$ from $\mathbb{R}_+ \times E \times E$ into $\mathbb{R}_+ \times E$, the statement d) becomes a corollary of c), and we have $k(x,z) = l(x,z)-x$ with $\Delta - x = 0$.

§3d. *Examples and comments*

Our aim is to give several examples of semimartingale Hunt processes, show the workings of our representation theorems in a few cases, and comment on the existence-uniqueness questions and the roles of the strong Markov property and quasi-left-continuity.

As we had mentioned earlier, every process with *stationary and independent increments* is an Itô process where the coefficient functions b, c, k are free of x and $k(x,z) = k_0(z)$ further satisfies $\int \nu(dz)(|k_0(z)|^2 \wedge 1) < \infty$, see (1.2). Ito processes with $k = 0$ are called *diffusions* (or quasi-diffusions by some authors), and they have been studied extensively. The following is a special diffusion process.

(3.55) ORNSTEIN-UHLENBECK PROCESSES. For the case of motions in $\mathbb{R}$, these are two-dimensional processes $X = (X^1, X^2)$ where X_t^1 is interpreted as the velocity and X_t^2 as the position of a particle moving in $\mathbb{R}$. Such a process X is an Itô process and satisfies (1.1) with

$$b(x_1,x_2) = (-\beta x_1, x_1), \quad c(x_1,x_2) = \begin{bmatrix} \sigma & 0 \\ 0 & 0 \end{bmatrix}, \quad k = 0,$$

where β and σ are some positive constants. In this case it is possible to solve (1.1) for X explicitly:

$$X_t^1 = e^{-\beta t} X_0^1 + \int_0^t e^{-\beta(t-s)}\, dW_s, \quad X_t^2 = \int_0^t X_s^1\, ds.$$

For each t, (X_t^1, X_t^2) has a Gaussian distribution.

The following are Itô process whose paths are of finite variation.

(3.56) STORAGE PROCESSES. These are Itô processes on $\mathbb{R}_+$ satisfying (1.5) with $\hat{b}(x) = -r(x)$ and $k(x,z) = k_0(z)$, where r is an increasing positive left-continuous function on $\mathbb{R}_+$ with $r(0) = 0$ and where $k_0 \geq 0$ satisfies $\int \nu(dz)\,(k_0(z) \wedge 1) < \infty$. Then, (1.5) can be re-written as

$$X_t = X_0 + Y_t - \int_0^t r(X_s)\,ds,$$

where (Y_t) is an increasing process with stationary and independent increments. One interprets X_t as the content of a storage system at time t; then Y_t becomes the cumulative input into the system during $(0,t]$, and $r(x)$ is the rate of release (per unit time) when the content is x. Under the conditions mentioned for k_0 and r, this equation has one and only one solution.

(3.57) REGULAR STEP PROCESSES. We will describe these constructively, in the case where the state space is $\mathbb{R}$, to provide an example of the workings of our representations. For a regular step process in the sense of [2] (they are always minimal), the probability law of the process is completely specified by its Lévy kernel L (chosen so that the Lévy system is (t,L)). One has $\lambda(x) = L(x,\mathbb{R}) < \infty$ for every x. If $\lambda(x) = 0$ then x is absorbing; otherwise, if $\lambda(x) > 0$ then $Q(x,dy) = L(x,dy)/\lambda(x)$ is the distribution of the position to be occupied after x.

We fix the measure ν on $\mathbb{R}$ to be the Lebesgue measure on $[0,\infty)$ and set $\nu((-\infty,0]) = 0$. The mapping $y \to L(x,(-\infty,y])$ is a right continuous increasing function with limit $\lambda(x)$ as $y \to \infty$. Let

$z \to l(x,z)$ be its right continuous functional inverse for $0 < z \le \lambda(x)$, and set

$$k(x,z) = \begin{cases} l(x,z)-x & \text{if } 0 < z < \lambda(x), \\ 0 & \text{if } z \ge \lambda(x). \end{cases}$$

Then, for any Borel set B not including 0, we have $L(x,B) = \int \nu(dz) \cdot 1_B(x+k(x,z))$.

Let N be a Poisson random measure on $\mathbb{R}_+ \times \mathbb{R}_+$ with mean measure $dt \cdot dz$, and consider the equation

$$X_t = X_0 + \int_0^t \int_0^\infty N(ds,dz)\, k(X_{s-},z).$$

If $N(\omega,\{s\}\times\{z\}) = 1$ for some (ω,s,z), then the path $X(\omega)$ jumps by the amount $k(x,z)$ if $X_{s-}(\omega) = x$ (with the obvious interpretation that $k(x,z) = 0$ means there is no jump; in particular, if $z \ge \lambda(x)$ then $k(x,z) = 0$ and there is no jump). So, we may describe the path $X(\omega)$ as follows. Starting at x, the path stays at x until the first time t where $N(\omega,\{t\}\times[0,\lambda(x)]) = 1$. At that time, say $T_1(\omega)$, the path jumps to $y = x + k(x,z)$ if $N(\omega;\{T_1(\omega)\}\times\{z\}) = 1$. Then, the path stays at y until the first time $t > T_1(\omega)$ such that $N(\omega,\{t\}\times[0,\lambda(y)]) = 1$; at that time, say $T_2(\omega)$, it jumps from y to $y+k(y,z)$ if $N(\omega,\{T_2(\omega)\}\times\{z\}) = 1$. And so on.

Thus, the atoms of the measure $N(\omega,\cdot)$ over the curve $t \to \lambda(X_t(\omega))$ play no role in constructing $X(\omega)$. When we reverse the problem and try to construct a Poisson random measure $\tilde{N}(\omega,\cdot)$ from the path $X(\omega)$, the atoms of $\tilde{N}(\omega,\cdot)$ above $t \to \lambda(X_t(\omega))$ are supplied by an auxiliary Poisson random measure constructed on a separate probability space $(\Omega',\mathcal{H}',P')$.

The following is a semimartingale Hunt process but is not an Itô process. It illustrates the important role played by the time change in Theorem (3.35) in smoothing the paths over the boundary.

(3.58) ABSOLUTE VALUE OF BROWNIAN MOTION. Let (X_t) be a standard Brownian motion on $\mathbb{R}$ (then $X_t - X_0$ is a Wiener process), and let $Y_t = |X_t|$. Then, Y is a semimartingale Hunt process. It follows from a result due to TANAKA that

$$Y_t = Y_0 + \int_0^t \operatorname{sgn}(X_s)\, dX_s + B_t$$

where (B_t) is the local time at 0 (for both X and Y). This shows that, in the decomposition (2.41) we have $Y^d = Y^e = 0$ and $Y^b = B$ and $Y^c = \int \operatorname{sgn} X\, dX$. We thus have $C_t = \langle Y^c, Y^c \rangle_t = t$. Now let

$$A_t = B_t + C_t = B_t + t.$$

Then,

$$B_t = \int_0^t b(Y_s) dA_s, \qquad t = \int_0^t c(Y_s)^2\, dA_s$$

with

$$b(x) = 1_{\{0\}}(x), \qquad c(x) = 1_{(0,\infty)}(x).$$

Let $\hat{A}$ be the functional inverse of A, and define

$$\hat{Y}_t = Y_{\hat{A}_t}.$$

We then have $\hat{Y}_t = \hat{Y}_0 + B(\hat{A}_t) + Y^c(\hat{A}_t)$. Clearly,

$$\hat{A}_t = \int_0^t 1_{(0,\infty)}(\hat{Y}_s)ds, \qquad B(\hat{A}_t) = \int_0^t 1_{\{0\}}(\hat{Y}_s)\, ds.$$

Now let W be an auxiliary Wiener process (defined on a separate probability space (Ω', H', P'), and set

$$d\tilde{W}_t(\omega,\omega') = 1_{(0,\infty)}(\hat{Y}_t(\omega))d\hat{Y}^c_t(\omega) + 1_{\{0\}}(\hat{Y}_t(\omega))dW_t(\omega').$$

Then, $\tilde{W}$ is a Wiener process on the enlarged space, and we have

$$\int_0^t 1_{(0,\infty)}(\hat{Y}_s)\, d\tilde{W}_s = \hat{Y}^c_t.$$

Thus, we have the representation

$$\hat{Y}_t = \hat{Y}_0 + \int_0^t 1_{\{0\}}(\hat{Y}_s)ds + \int_0^t 1_{(0,\infty)}(\hat{Y}_s)d\tilde{W}_s, \tag{3.59}$$

which shows that $\tilde{Y}$ is a continuous Itô process. Further, $Y_t = \hat{Y}(A_t)$, and A is the functional inverse of $\hat{A}$.

This example illustrates the role of the time change in smoothing the paths of Y at the boundary point 0. The effect of the time change is to dilate the time set $\{ t: Y_t = 0 \}$ so that its Lebesgue measure becomes positive, but the excursions away from 0 are not altered at all. This yields the process $\hat{Y}$. Working with $\hat{Y}$, the excursions of $\hat{Y}$ outside 0 can be used to define the excursions of a Wiener process $\tilde{W}$ outside 0, but the behavior of $\tilde{W}$ at 0 needs to be supplied separately, by using an auxiliary process W.

Unfortunately, starting with a given $\tilde{W}$, the equation (3.59) has a large number of solutions (for example, $\tilde{W}$ is a solution that spends no time at the boundary).

(3.60) EXISTENCE AND UNIQUENESS QUESTIONS. We had shown in Theorem (3.35) that, if a given Hunt process is also a semimartingale, and if it admits (t,b,c,K) as a system of local characteristics, then it satisfies (1.1). Suppose that the state space is $\mathbb{R}$ and that X is continuous. If we go over the way b and c are defined, we see that the only necessary property they have is Borel measurability and they can be assumed to be bounded (the latter is because of the time change). In the converse direction, if b and c are Lipschitz continuous, then (1.1) with $k = 0$ has a unique solution. More generally, STROOCK and VARADHAN [24] show that if b is bounded and measurable and c is bounded and continuous, then it is possible to construct a probability space and a process X on it such that X is an Itô process, and therefore satisfies (1.1) on some extension of the original space.

Thus, assuming b and c to be bounded and measurable as is necessary by our results, there remains a gap in characterizing c for which (1.1) can be solved at least in the weak sense (of existence of a probability space such that ...). The gap is between the necessary condition of measurability for c and the known sufficient condition of continuity [24].

Our results say nothing about uniqueness of solutions to (1.1): given an Ito process, it satisfies (1.1) for some b, c, k, but there may be other solutions. Generally, the questions of uniqueness have to do with the behavior on the essential boundary of the state space. We have seen an instance of it in (3.59). The following is another.

(3.61) EXAMPLE. Let $(x_n) \subset (0,\infty)$ be a strictly decreasing sequence with limit 0, let $(y_n) \subset (-\infty,0)$ be a strictly increasing sequence with limit 0, and put $E = \{ x_n: n \geq 0 \} \cup \{ y_n: n \geq 0 \} \cup \{0\}$. Let X be a process with state space E, and such that every one of the points x_n and y_n are holding points, x_0 and y_0 are absorbing. If X starts at x_n then the path goes through $x_n, x_{n-1}, \ldots, x_0$ in that order. If X starts at y_n, then it goes through $y_n, y_{n-1}, \ldots, y_0$ in that order. This rough description can be made precise, and the description of the probability measures $\mathbb{P}_x$ for $x \in E\backslash\{0\}$ presents no difficulty.

Also, there is no difficulty in constructing a Poisson random measure $\tilde{N}$ and choosing a function k so that

$$(3.62) \qquad X_t(\omega) = X_0(\omega) + \int_0^t \int_{-\infty}^{\infty} \tilde{N}(\omega,\omega';ds,dz)\, k(X_{s-}(\omega),z)$$

for all ω and ω', even if $X_0(\omega) = 0$. Moreover, if $X_0(\omega) \neq 0$, this equation has exactly one solution. But if $X_0(\omega) = 0$, the equation has exactly two solutions (one increasing over the x_n and the other decreasing over the y_n).

Given the equation (3.62) with the proper k, we cannot tell if X is strong Markov with respect to (F^o_{t+}). If it is known that X is strong Markov relative to (F^o_{t+}), then the zero-one law guarantees that

$$\mathbb{P}_0\{\, X_t \in \{\, x_n : n \geq 0 \,\} \quad \text{for all } t > 0 \,\} = 0 \quad \text{or} \quad 1,$$

but we cannot say anything further.

The problems of non-uniqueness of this type will ever be with us. This example, and most of the known cases of non-uniqueness, are really trivial in nature: take two open sets C and D whose boundaries touch each other at a point δ, set $E = C \cup D \cup \{\delta\}$, and demand that the process live either in C or in D essentially. Then, we will have the difficulty of deciding what to do at δ.

(3.63) ROLE OF STRONG MARKOV PROPERTY. Given a semimartingale Markov process X, our representation theorem may fail if X is not strong Markov. Recall that every increasing continuous strong Markov process is deterministic except in the choice of the initial state. The following example, due to LEVY, is a continuous increasing *non-deterministic* Markov process that is not strong Markov and for which our theorems do not hold.

(3.64) EXAMPLE. For rational $r \in (0,1]$ pick $\lambda(r)$ so that $\sum \lambda(r)^{-1} < \infty$. For rationals $r > 1$ pick $\lambda(r) = \lambda(r-n)$ if $n < r \leq n+1$, $n = 1,2,\ldots$. For each rational $r > 0$, let Z_r be an exponentially distributed random variable with mean $1/\lambda(r)$, and suppose the Z_r are mutually independent. Let $S_0 = 0$,

$$S_x = \sum_{0<r\leq x} Z_r, \quad x > 0,$$

and let

$$X_t = \inf\{\ x: S_x > t\ \}, \quad t \geq 0.$$

Since $(S_x)_{x\geq 0}$ is right continuous and strictly increasing with limit ∞, the process X is increasing and continuous. It is clear that X is Markov, each rational r is a holding point, and X spends no time on the set of irrationals. None of our representations work for this X.

(3.65) ROLE OF QUASI-LEFT-CONTINUITY. In case X has jumps, its quasi-left-continuity is essential for the conversion of its jump measure to a Poisson random measure. The following is an example of a Hunt process that is not quasi-left-continuous, and for which our theorems do not work.

(3.66) EXAMPLE. Let $t_0 = 0 < t_1 < t_2 < \cdots$ be a fixed sequence of times increasing to ∞, and set $X_t = t_{n+1}-t$ if $t_n \le t < t_{n+1}$. This process satisfies all the conditions for a Hunt process except the quasi-left-continuity. There is no way of accounting for its jumps by a Poisson random measure.

§4. Proof of The Fundamental Result

A representation like (3.8) is well-known for continuous local martingales without our Markovian setting. For integer-valued random measures, again without the Markovian setting, a representation like (3.9) was shown to hold by GRIGELIONIS [12]; see also EL KAROUI and LEPELTIER [7], or JACOD [16], or KABANOV, LIPTSER and SHIRYAYEV [17]. Even in that simpler case, with only one probability measure, the proof is quite difficult. In our setting, we require freedom from x in decompositions and constructions with respect to $\mathbb{P}_x$, and we need the strong Markov property over the enlargements. For these reasons, we cannot use the proofs of [7], [12], and [17], even though the constructions involved are quite similar. Our aim in this section is to give the proof in its entirety for the fundamental result, namely, Theorem (3.7).

In order to prove (3.7) it is sufficient to prove it in the following two extreme cases:

i) $Y = (Y^i)_{i\in I}$ satisfies (3.1) and $\Gamma = 0$;

ii) Γ satisfies (3.2) and $Y = 0$.

To see this, suppose (3.7) holds under both i) and ii), and let Y and Γ satisfy (3.1) and (3.2) respectively. Then, there is a strong Markov extension $(\hat{X},\hat{H}_t)$ of (X,H_t) supporting a Poisson random measure $\tilde{N}$ such that Γ satisfies (3.9). Now Y satisfies (3.1) relative to

$(\hat{X},\hat{H}_t)$ by Proposition (2.50). Thus, since (3.7) is assumed to hold in the case i), there is a strong Markov extension $(\tilde{X},\tilde{H}_t)$ of $(\hat{X},\hat{H}_t)$ supporting a Wiener process $\tilde{W}$ such that (3.8) holds. Note that $(\tilde{X},\tilde{H}_t)$ is a strong Markov extension of (X,H_t) by Proposition (2.49), and $\tilde{N}$ is a Poisson random measure over $(\tilde{X},\tilde{H}_t)$ by Corollary (2.52). Thus, Theorem (3.7) holds in the general case where both Y and Γ are non-trivial.

In §4a below we will prove (3.7) in the first extreme case, and then in §§4b - 4e we will show (3.7) for the second case.

§4a. *Representation of additive continuous local martingales*

The idea behind the proof is quite simple. Let Y be an additive continuous local martingale on $\mathbb{R}$ satisfying (3.1). Then c is scalar valued. If c never vanishes, we put $\tilde{W}_t = \int_0^t c(X_s)^{-1}\, dY_s$, and then it is immediate that $\tilde{W}$ is a Wiener process over (X,H_t) and that $Y_t = \int_0^t c(X_s)\, d\tilde{W}_s$. However, if $c(x)$ vanishes for some x, it is impossible to determine $d\tilde{W}_t$ directly from dY_t when $c(X_t) = 0$. Then, the idea is to put

$$d\tilde{W}_t = \frac{1}{c(X_t)} I_{\{c(X_t)\neq 0\}}\, dY_t + I_{\{c(X_t)=0\}}\, dW_t,$$

where W is an auxiliary Wiener process independent of X, and then check that $\tilde{W}$ is a Wiener process.

From here on, $Y = (Y^i)_{i\in I}$ satisfies (3.1), and c is as described there.

We first enlarge the probability spaces $(\Omega,H,\mathbb{P}_x)$. For that purpose, let $(\Omega',H',H'_t,\theta'_t,W_t,P')$ be a canonical Wiener process indexed by I; in other words, Ω' is the set of all continuous paths $w\colon \mathbb{R}_+ \to \mathbb{R}^I$ with $w(0) = 0$, $W_t(w) = w(t)$, $\theta'_t w = w(t+\cdot)-w(t)$,

$H'_t = \sigma(W_s;\ s \le t)$, $H' = \vee_t H'_t$, and P' is the unique probability measure under which the W^i are independent standard (one-dimensional) Wiener processes. We set

$$(4.1)\qquad (\tilde{\Omega},\tilde{H},\tilde{\mathbb{P}}_x) = (\Omega,H,\mathbb{P}_x) \times (\Omega',H',P'),$$

$$(4.2)\qquad \tilde{\theta}_t(\omega,w) = (\theta_t\omega,\theta'_t w), \qquad \tilde{H}_t = \bigcap_{u>t} H_u \otimes H'_u ,$$

and as usual we denote by the same symbol any function on Ω or Ω' and its natural extension to $\tilde{\Omega}$.

(4.3) PROPOSITION. Let $\tilde{X} = (\tilde{\Omega},\tilde{H},\tilde{H}_t,\tilde{\theta}_t,X_t,\tilde{\mathbb{P}}_x)$. Then, $(\tilde{X},\tilde{H}_t)$ is a strong Markov extension of (X,H_t), and W is a Wiener process over $(\tilde{X},\tilde{H}_t)$.

PROOF. The only condition that is not totally obvious is (2.47ii), and it is sufficient to demonstrate it for $Z(\omega,w) = V(\omega)V'(w)$ with $V \in bH$ and $V' \in bH'$. Let T be a finite stopping time of (H_t), and let $\tilde{U} \in b\tilde{H}_T$. Then, in view of (4.1) and (4.2),

$$(4.4)\qquad \tilde{\mathbb{E}}_x[\,\tilde{U}\, Z\circ\tilde{\theta}_T\,] = \int \mathbb{P}_x(d\omega) \int P'(dw)\tilde{U}(\omega,w)V(\theta_{T(\omega,w)}\omega)V'(\theta'_{T(\omega,w)}w)$$

$$= \int \mathbb{P}_x(d\omega) \int P'(dw)\tilde{U}(\omega,w)V(\theta_{T(\omega,w)}\omega)\ E'[V']$$

by the strong Markov property of the Wiener process $(\Omega',H',H'_{t+},\theta'_t,W_t,P')$, since for fixed ω, $T(\omega,\cdot)$ is a stopping time of (H'_{t+}), $\tilde{U}(\omega,\cdot) \in bH'_{T(\omega,\cdot)+}$, $V(\theta_{T(\omega,\cdot)}\omega) \in bH'_{T(\omega,\cdot)+}$, and $V' \in bH'$.

On the other hand, for fixed $w \in \Omega'$, $T(\cdot,w)$ is a stopping time of (H_t), $\tilde{U}(\cdot,w) \in bH_{T(\cdot,w)}$, and $V \in bH$. Thus, since (H_t) is a strong

Markov filtration for X, the last member of (4.4) is equal to

$$(4.5) \qquad \int P'(dw) \int \mathbb{P}_x(d\omega)\, \tilde{U}(\omega,w)\, \mathbb{E}_{X_{T(\omega,w)}(\omega)}[V]\, E'[V']$$

$$= \tilde{\mathbb{E}}_x[\, \tilde{U}\, \tilde{\mathbb{E}}_{X_T}[Z]\,]$$

as required.

Now consider the additive continuous local martingales M^i, $i \in I$, of $(X, \mathcal{H}_t)$ constructed in Theorem (2.16) and satisfying (2.17), (2.18) with $a_{ij} = c_{ij}$, and (2.19). Because of Proposition (2.50), these properties remain valid over $(\tilde{X}, \tilde{\mathcal{H}}_t)$, the extension defined by (4.1) and (4.2). Recalling the Wiener process W in (4.3), we now define

$$(4.6) \qquad \tilde{W}^i_t = M^i_t + \int_0^t (1 - 1_{B_i}(X_s))\, dW^i_s, \qquad i \in I,$$

on the space $(\tilde{\Omega}, \tilde{\mathcal{H}}, \tilde{\mathcal{H}}_t, \tilde{\mathbb{P}}_x)$. Since the stochastic integral defining M^i does not depend on which $\mathbb{P}_x$ is used, $\tilde{W}^i$ is the same under every $\tilde{\mathbb{P}}_x$. The following shows that $\tilde{W}$ is the Wiener process we were seeking.

(4.7) PROPOSITION. The process $\tilde{W}$ is a Wiener process over $(\tilde{X}, \tilde{\mathcal{H}}_t)$ and the equalities (3.8) hold.

PROOF. It is clear that $\tilde{W}$ is an additive continuous local martingale over $(\tilde{X}, \tilde{\mathcal{H}}_t)$. Since M and W are independent, (2.17), (2.19), and (4.6) yield

$$\langle \tilde{W}^i, \tilde{W}^j \rangle_t = \int_0^t 1_{B_i}(X_s)\delta_{ij}\, ds + \int_0^t (1 - 1_{B_i}(X_s))\delta_{ij}\, ds = \delta_{ij}\, t.$$

Hence, $\tilde{W}$ is a Wiener process by Proposition (2.24). Moreover, we have $c_{ik}(1 - 1_{B_k}) = 0$ for all $i, k \in I$, which together with (4.6) and (2.18) imply that (3.8) holds.

(4.8) REMARK. If $c_{ii}(x)$ differs from 0 for all i and x, then we have $B_i = E$ in Lemma (4.5), and $M = (M^i)$ is a Wiener process on (X, H_t). In that case, we put $\tilde{W} = M$, we have no use for the auxiliary Wiener process W, and hence, the extension $(\tilde{X}, \tilde{H}_t)$ is not needed.

§4b. *Representation of random measures: Outline*

In order to gain a rough idea of what is to be done, suppose Theorem (3.7) holds, and consider the problem of constructing $\tilde{N}$. Then, we know the measure Γ, and we want to deduce $\tilde{N}$ satisfying (3.9). Recall that Γ satisfies (2.25); so, we may define a D_Δ-valued process β by setting $\beta_t = \Delta$ if $\Gamma(\{t\}\times D) = 0$ and $\beta_t = y$ if $\Gamma(\{t\}\times\cdot) = \varepsilon_y(\cdot)$. Now, fix $t \in \mathbb{R}_+$, $\omega \in \Omega$, $\omega' \in \Omega'$, and set $x = X_{t-}(\omega)$, $y = \beta_t(\omega)$. There are three cases:

a) If $y \neq \Delta$ and $k(x,z) = y$ for exactly one z, say $z = \hat{k}(x,y)$, then $\tilde{N}((\omega,\omega');\{t\}\times B) = 1_B \circ \hat{k}(x,y)$, which does not depend on ω' at all.

b) If $y \neq \Delta$ but $k(x,z) = y$ for all z in some non-singleton set B_{xy}, then we know that $\tilde{N}((\omega,\omega');\{t\}\times B_{xy}) = 1$. Given this information, the actual location of the corresponding atom is some point $z_0(\omega')$, where $\omega' \to z_0(\omega')$ has the distribution $\nu(dz)/\nu(B_{xy})$ on the set B_{xy}. It follows that

$$\tilde{N}((\omega,\omega');\{t\}\times B) = 1_B \circ \hat{k}(x,y,U(\omega')) \tag{4.9}$$

where $\omega' \to U(\omega')$ has a uniform distribution on $(0,1]$ and where $\hat{k}$ is selected so that $\hat{k}(x,y,U)$ has the distribution $\nu(\cdot)/\nu(B_{xy})$ on the set B_{xy}.

c) If $y = \Delta$ but $B_x = \{ z: k(x,z) = \Delta \}$ is not empty, we have no information on $\tilde{N}((\omega,\omega');\{t\}\times\cdot)$, and then we set

$$\tilde{N}((\omega,\omega');\{t\}\times B) = N(\omega';\{t\}\times(B\cap B_x)) \tag{4.10}$$

where N is some auxiliary Poisson random measure on Ω'.

Thus, for constructing $\tilde{N}$, we need an auxiliary Poisson random measure N and an auxiliary sequence (U_n) of independent uniformly distributed random variables on $(0,1]$. We will do the constructions associated with (U_n) in §4c, and construct N in §4d. Finally, $\tilde{N}$ will be constructed in §4e by the recipes (4.9) and (4.10).

§4c. *Adding uniform variables*

Throughout the remainder of this section, the setup is that of Theorem (3.7) with $Y = 0$. Since Γ satisfies (2.25), there exists a D_Δ-valued process β such that

$$\Gamma(\omega;dt,dy) = \sum_{s>0} I_{\{\beta_s(\omega)\in D\}}\, \varepsilon_{(s,\beta_s(\omega))}(dt,dy). \tag{4.11}$$

The assumptions on Γ imply that β is $(\mathcal{H}_t)$-optional and homogeneous in the sense that, for every t, the processes $\beta_{t+\cdot}$ and $\beta_\cdot\circ\theta_t$ are indistinguishable.

Let $(B_n)_{n\in\mathbb{N}^*}$ be the $\mathcal{D}$-measurable partition of D encountered in (2.28ii), and define

$$S_n = \inf\{ t: \beta_t \in B_n \}, \qquad n \in \mathbb{N}^*. \tag{4.12}$$

Each S_n is a terminal time, that is, $S_n = t + S_n \circ \theta_t$ almost surely on $\{ S_n > t \}$. Using the methods of WALSH [26], then, we can modify S_n on a null set to obtain a new stopping time (which we denote by S_n again) that is an exact terminal time (i.e. $S_n = t + S_n \circ \theta_t$ everywhere on $\{ S_n > t \}$), and (4.12) holds almost surely.

We let S_{np} be the p-th iterate of S_n: put $S_{n1} = S_n$ and

(4.13) $$S_{n,p+1} = S_{np} + S_n \circ \theta_{S_{np}}, \quad p \in \mathbb{N}^* .$$

Then,

$$J_n(t) = \sum_{p \in \mathbb{N}^*} 1_{(0,t]} \circ S_{np}, \quad t \geq 0,$$

is increasing and perfectly additive:

(4.15) $$J_n(t+u,\omega) = J_n(t,\omega) + J_n(u,\theta_t\omega)$$

for all t, u, and ω. Moreover, for all $t \geq 0$,

(4.16) $$J_n(t) = \Gamma((0,t]\times B_n) \quad \text{a.s.}$$

We are now ready for the first extension. Let

(4.17) $$(\Omega^1, \mathcal{H}^1, P^1) = ((0,1], \mathcal{B}(0,1], \text{Leb})^{\mathbb{N}^* \times \mathbb{N}^*},$$

and let $U_{np}(w)$ denote the (n,p)-coordinate of $w \in \Omega^1$. Obviously, the U_{np} are independent and uniformly distributed on $(0,1]$ as random variables on $(\Omega^1, \mathcal{H}^1, P^1)$. For $\omega \in \Omega$ fixed, let $\theta_t^\omega : \Omega^1 \to \Omega^1$ be the shift characterized by

(4.18) $$U_{np}(\theta_t^\omega w) = U_{n,J_n(t,\omega)+p}(w), \quad w \in \Omega^1,$$

and define an increasing family of σ-fields $(\mathcal{H}_t^\omega)_{t\geq 0}$ on Ω^1 by

(4.19) $$\mathcal{H}_t^\omega = \sigma\{\, U_{np}:\ n \in \mathbb{N}^*,\ p \leq J_n(t,\omega)\,\}.$$

(4.20) LEMMA. Let $\omega \in \Omega$ be fixed, and let $T\colon \Omega^1 \to [0,\infty]$ be a stopping time of $(\mathcal{H}_{t+}^\omega)$. Then, for every $W \in b\mathcal{H}_{T+}^\omega$ and $Z \in b\mathcal{H}^1$,

$$E^1[\, W\cdot Z\circ\theta_T^\omega \,] = E^1[W]\cdot E^1[Z].$$

PROOF. Let ω, T, W, Z be fixed. A monotone class argument enables us to assume that $Z = f\circ(U_{np}:\ n \in M,\ p \in \mathbb{N}^*)$ where $M \subset \mathbb{N}^*$ is finite and f is some bounded Borel function on $(0,1]^{M\times\mathbb{N}^*}$.

For every vector $j = (j_n)_{n\in M} \in \mathbb{N}^M$ we define an index set $\mathbb{K}_j$ by

$$\mathbb{K}_j = \{\, (n,p) \in \mathbb{N}^*\times\mathbb{N}^*:\ n \in M,\ p > j_n \,\},$$

and let

$$A_j = \{\, w \in \Omega^1:\ J_n(T(w),\omega) = j_n,\ n \in M \,\}.$$

Then, (4.18) implies that

(4.21) $$Z\circ\theta_T^\omega = f\circ(U_{np}:\ (n,p) \in \mathbb{K}_j\,) \quad \text{on } A_j,$$

and we will show below that

(4.22) $$W\, I_{A_j} = g_j\circ(U_{np}:\ (n,p) \notin \mathbb{K}_j\,)$$

for some bounded Borel function g_j on $(0,1]^{\mathbb{N}^* \times (\mathbb{N}^* \setminus \mathbb{K}_j)}$. Then, since the U_{np} are independent and identically distributed on (Ω^1, H^1, P^1),

$$E^1[\ W\ I_{A_j}\ Z\circ\theta_T^\omega\ ] = E^1[\ g_j\circ(U_{np}:(n,p)\notin \mathbb{K}_j\ )\ f\circ(U_{np}:(n,p)\in \mathbb{K}_j\ )\ ]$$

$$= E^1[\ g_j\circ(U_{np}:(n,p)\notin \mathbb{K}_j\ )\ ]\ E^1[\ f\circ(U_{np}:(n,p)\in \mathbb{K}_j\ )\ ]$$

$$= E^1[\ W\ I_{A_j}\ ]\ E^1[\ f\circ(U_{np}:(n,p)\in M\times\mathbb{N}^*\ )\ ]$$

$$= E^1[\ W\ I_{A_j}\ ]\ E^1[Z].$$

This yields the desired result when both sides are summed over all $j \in \mathbb{N}^M$.

To complete the proof, there remains to show (4.22). For this, it is enough to show that

$$W\ I_{A_j}(w) \ = \ W\ I_{A_j}(\hat{w}) \tag{4.23}$$

for all w and $\hat{w}$ in A_j such that $U_{np}(w) = U_{np}(\hat{w})$ for all $(n,p) \notin \mathbb{K}_j$. Let $w, \hat{w} \in A_j$. Because J_n is right-continuous and integer-valued for every n, because $J_n(T(w),\omega) = J_n(T(\hat{w}),\omega) = j_n$ for all $n \in M$, and since M is finite, there exists $t > T(\omega)\vee T(\hat{w})$ such that

$$J_n(t,\omega) \ = \ j_n, \quad n \in M. \tag{4.24}$$

But, $W\ I_{A_j}\ I_{\{T<t\}} \in bH_t^\omega$, which implies by definition (4.19) of H_t^ω that $W\ I_{A_j}\ I_{\{T<t\}}$ is a function of $(U_{np}:\ n \in \mathbb{N}^*,\ p \le J_n(t,\omega))$. Thus, by

(4.24),

$$W\, I_{A_j}(w) = W\, I_{A_j}\, I_{\{T<t\}}(w)$$

$$= W\, I_{A_j}\, I_{\{T<t\}}(\hat{w}) = W\, I_{A_j}(\hat{w}),$$

which is the desired result (4.23).

We now proceed to the first enlargement of $(X,\mathcal{H}_t)$. We put

(4.25) $$(\bar{\Omega},\bar{\mathcal{H}},\bar{\mathbb{P}}_x) = (\Omega,\mathcal{H},\mathbb{P}_x) \times (\Omega^1,\mathcal{H}^1,P^1),$$

(4.26) $$\bar{\theta}_t(\omega,w) = (\theta_t\omega, \theta_t^\omega w),$$

(4.27) $$\bar{\mathcal{H}}_t^o = \{\, A \in \mathcal{H}_t \otimes \mathcal{H}^1 :\ I_A(\omega,\cdot) \in \mathcal{H}_t^\omega \ \text{ for every } \ \omega \in \Omega \,\}$$

$$\bar{\mathcal{H}}_t = \cap_{s>t}\, \bar{\mathcal{H}}_s^o .$$

For functions defined on Ω or Ω^1, we let the same symbol denote their natural extensions to $\bar{\Omega}$.

(4.28) PROPOSITION. Let $\bar{X} = (\bar{\Omega},\bar{\mathcal{H}},\bar{\mathcal{H}}_t,\bar{\theta}_t,X_t,\bar{P}_x)$. Then, $(\bar{X},\bar{\mathcal{H}}_t)$ is a strong Markov extension of $(X,\mathcal{H}_t)$.

PROOF. The only conditions that need checking concern $(\bar{\theta}_t)$ and the strong Markov property (2.47ii). Since (4.15) holds identically, it follows from the definition (4.18) of θ_t^ω that $\theta_{t+s}^\omega w = \theta_t^{\theta_s\omega}(\theta_s^\omega w)$, from which we obtain $\bar{\theta}_{t+s} = \bar{\theta}_t \circ \bar{\theta}_s$. It is easy to check that $(t,\bar{\omega}) \to \bar{\theta}_t\bar{\omega}$ is a measurable mapping from $([0,s] \times \bar{\Omega},\ \mathcal{B}[0,s] \otimes \bar{\mathcal{H}}_{s+u})$ into $(\bar{\Omega},\bar{\mathcal{H}}_u)$, and thus the measurability condition in (2.47ii) is satisfied.

To prove the extended strong Markov property (2.48), let T be a finite stopping time of $(\bar{H}_t)$, and let $W \in b\bar{H}_T$ and $Z \in b\bar{H}$. We want to show that

$$\bar{\mathbb{E}}_x[\, W\, Z\circ\bar{\theta}_T \,] = \bar{\mathbb{E}}_x[\, W\, \bar{\mathbb{E}}_{X_T}[Z]\,]. \tag{4.29}$$

By the monotone class theorem, it is sufficient to show this by assuming that $Z(\omega,w) = Z_0(\omega)Z_1(w)$ for some $Z_0 \in bH$ and $Z_1 \in bH^1$.

For fixed $\omega \in \Omega$, by the definition (4.27) of $\bar{H}_t$, $T(\omega,\cdot)$ is a stopping time of (H^{ω}_{t+}), $W(\omega,\cdot) \in bH^{\omega}_{T(\omega,\cdot)+}$, and $Z_0(\theta_{T(\omega,\cdot)}\omega) \in bH_{T(\omega,\cdot)+}$. Thus, by Lemma (4.20),

$$\begin{aligned}\bar{\mathbb{E}}_x[\, W\, Z\circ\bar{\theta}_T \,] &= \int \mathbb{P}_x(d\omega) \int P^1(dw) W(\omega,w) Z_0(\theta_{T(\omega,w)}\omega) Z_1(\theta_{T(\omega,w)}w) \\ &= \int \mathbb{P}_x(d\omega) \int P^1(dw) W(\omega,w) Z_0(\theta_{T(\omega,w)}\omega)\, E^1[Z_1].\end{aligned} \tag{4.30}$$

On the other hand, for fixed $w \in \Omega^1$, $T(\cdot,w)$ is a stopping time of (H_t), $W(\cdot,w) \in bH_{T(\cdot,w)}$, and $Z_0 \in bH$. Thus, since (H_t) is a strong Markov filtration of X, the last member of (4.30) is equal to

$$\begin{aligned}&\int P^1(dw) \int \mathbb{P}_x(d\omega)\, W(\omega,w)\, \mathbb{E}_{X_{T(\omega,w)}(\omega)}[Z_0]\, E^1[Z_1] \\ &\qquad = \int \mathbb{P}_x(d\omega) \int P^1(dw)\, W(\omega,w)\, \bar{\mathbb{E}}_{X_{T(\omega,w)}(\omega)}[Z] \\ &\qquad = \bar{\mathbb{E}}_x[\, W\, \bar{\mathbb{E}}_{X_T}[Z]\,],\end{aligned}$$

which is as desired.

§4d. *Adding the auxiliary Poisson random measure*

This is the final enlargement. Let $(\Omega^2, H^2, H^2_t, \theta^2_t, N, P^2)$ be a canonical Poisson random measure on $\mathbb{R}_+ \times \mathbb{R}$ with mean measure $dt\,\nu(dz)$: each $w \in \Omega^2$ is a sequence $w = (t_n, z_n)_{n \in \mathbb{N}} \subset \mathbb{R}_+ \times \mathbb{R}$ with $\sup t_n = \infty$,

$$N(w,\cdot) = \Sigma_n\, \varepsilon_{(t_n,z_n)}(\cdot), \quad H^2_t = \sigma\{N(B) : B \subset [0,t] \times \mathbb{R} \text{ Borel}\}, \quad H^2 = \vee_t H^2_t$$

and the shifts θ_t are such that $N(\theta^2_t w; ds, dz) = N(w; ds-t, dz)\, 1_{[t,\infty)}(s)$. The construction of such things is well-known. We set

$$(\tilde{\Omega}, \tilde{H}, \tilde{\mathbb{P}}_x) = (\bar{\Omega}, \bar{H}, \bar{\mathbb{P}}_x) \times (\Omega^2, H^2, P^2)$$

$$\tilde{\theta}_t(\bar{\omega}, w) = (\bar{\theta}_t \bar{\omega}, \theta^2_t w), \qquad \tilde{H}_t = \cap_{s>t}\, \bar{H}_s \otimes H^2_s .$$

The proof of the following is exactly the same as that of (4.3) and will thus be omitted.

(4.31) PROPOSITION. Let $\tilde{X} = (\tilde{\Omega}, \tilde{H}, \tilde{H}_t, \tilde{\theta}_t, X_t, \tilde{P}_x)$. Then, $(\tilde{X}, \tilde{H}_t)$ is a strong Markov extension of $(\bar{X}, \bar{H}_t)$, and therefore of (X, H_t) by Proposition (2.49). The process N is a Poisson random measure over $(\tilde{X}, \tilde{H}_t)$ with mean measure $dt\,\nu(dz)$.

§4e. *Construction of the Poisson random measure* $\tilde{N}$

Our aim is to construct $\tilde{N}$ figuring in (3.9). We start with the following computational lemma, which provides insight into the representation (3.5) for the kernel K, and in fact proves (3.4) as a by-product. We give a full proof for reasons of completeness, and also because it might prove useful in applications.

(4.32) LEMMA. Let K be a positive kernel from $(E,\mathcal{E}_0)$ into $(D,\mathcal{D})$ satisfying (2.32).

a) There exists an $\mathcal{E}_0 \otimes \mathcal{D} \otimes \mathcal{B}(0,1]$ measurable function $\tilde{k}$: $E \times D \times (0,1] \to \mathbb{R}$ and an $\mathcal{E}_0 \otimes \mathcal{R}$ measurable function $k: E \times \mathbb{R} \to D_\Delta$ such that

$$(4.33) \qquad \int_0^1 du \int_A K(x,dy)\, 1_B \circ \tilde{k}(x,y,u) = \int_B \nu(dz)\, 1_A \circ k(x,z)$$

for every $A \in \mathcal{D}$, $B \in \mathcal{R}$, and $x \in E$. Moreover, for every $x \in E$ and $K(x,dy) \times du$ almost every $(y,u) \in D \times (0,1]$,

$$(4.34) \qquad k(x,\tilde{k}(x,y,u)) = y.$$

b) If $K(x,\cdot)$ is diffuse, then we can choose $\tilde{k}(x,y,u)$ to be free of u.

PROOF. *Step 1.* Let γ be a finite positive kernel from $(E,\mathcal{E}_0)$ into $([0,1],\mathcal{B}[0,1])$, and let a,b: $(E,\mathcal{E}_0) \to (\mathbb{R},\mathcal{R})$ be such that $\gamma(x,[0,1]) = \nu((a(x),b(x)])$ for every x (since ν is diffuse and infinite, this is possible). We set

$$\hat{g}(x,y) = \inf\{\, z: \nu((-\infty,z] \cap (a(x),b(x)]) > \gamma(x,[0,y])\,\},\ y \in [0,1],$$

$$g(x,z) = \inf\{\, y: \gamma(x,[0,y]) > \nu((-\infty,z] \cap (a(x),b(x)])\,\},\quad z \in \mathbb{R},$$

with $g(x,z) = 1$ in the last formula if $\{\cdots\}$ is empty. If every $\gamma(x,\cdot)$ is diffuse, it is easy to check that

$$(4.35) \quad \int_A \gamma(x,dy)\, 1_B \circ \hat{g}(x,y) = \int_{B \cap (a(x),b(x)]} \nu(dz)\, 1_A \circ g(x,z)$$

for all $A \in \mathcal{B}[0,1]$, $B \in \mathcal{R}$, $x \in E$.

Step 2. Let $\hat{f}: D \to C \subset [0,1]$ be a bi-measurable bijection from D into a Borel subset C of $[0,1]$, and let $f: [0,1] \to D_\Delta$ be its inverse, with $f(t) = \Delta$ if $t \in [0,1]\setminus C$.

Assume $K(x,\cdot)$ is diffuse and finite for every $x \in E$, and let $\gamma(x,\cdot) = K(x,f^{-1}(\cdot))$. Then, γ satisfies the hypotheses of step 1. Let a, b, g, $\hat{g}$ be picked as in Step 1, and set $h(x,z) = f \circ g(x,z)$ and $\hat{h}(x,y) = \hat{g}(x,\hat{f}(y))$. Then, (4.35) yields

$$(4.36)\quad \int_A K(x,dy)\, 1_B \circ \hat{h}(x,y) = \int_{B \cap (a(x),b(x)]} \nu(dz)\, 1_A \circ h(x,z)$$

for all $A \in \mathcal{D}$, $B \in \mathcal{R}$, and $x \in E$.

Step 3. Assume $K(x,\cdot)$ is diffuse for every x and suppose (2.32) holds. We can find an $\mathcal{E}_0 \otimes \mathcal{D}$-measurable partition (F_n) of $E \times D$ such that $K(x,F_n^x) < \infty$ for all $x \in E$ and n, where $F_n^x = \{ y: (x,y) \in F_n \}$. Since ν is diffuse and infinite and σ-finite, we can find functions $a_n, b_n: (E,\mathcal{E}_0) \to (\mathbb{R},\mathcal{R})$ such that the intervals $G_n^x = (a_n(x),b_n(x)]$ are disjoint and satisfy $\nu(G_n^x) = K(x,F_n^x)$. Now, the kernel $K_n(x,dy) = K(x,dy)1_{F_n}(x,y)$ satisfy the hypotheses of Step 2. Pick $\hat{h}_n$ and h_n as in Step 2 relative to $K_n(x,dy)$ and define

$$\hat{k}(x,y) = \hat{h}_n(x,y) \qquad \text{if } (x,y) \in F_n,$$

$$k(x,z) = \begin{cases} h_n(x,z) & \text{if } z \in G_n^x, \\ \Delta & \text{if } z \notin \bigcup_n G_n^x. \end{cases}$$

Since $K = \Sigma_n K_n$, summing the formula (4.36) written for K_n over all n, we obtain

$$(4.37) \qquad \int_A K(x,dy)\ 1_B\circ\hat{k}(x,y) = \int_B \nu(dz)\ 1_A\circ k(x,z)$$

for all $A \in \mathcal{D}$, $B \in \mathcal{R}$, $x \in E$. This completes the proof of part b) of the Lemma (take $\tilde{k}(x,y,u) = \hat{k}(x,y)$).

Step 4. To prove a), we apply the results of Step 3 to the diffuse measure $K(x,dy)du$ on $D \times (0,1]$. This yields measurable functions $\tilde{k}\colon E \times D \times (0,1] \to \mathbb{R}$ and $k^*\colon E \times \mathbb{R} \to (D\times(0,1])\cup\{\Delta\}$ such that

$$\int_C du \int_A K(x,dy)\ 1_B\circ\tilde{k}(x,y,u) = \int_B \nu(dz)\ 1_{A\times C}\circ k^*(x,z)$$

for all $A \in \mathcal{D}$, $B \in \mathcal{R}$, $C \in \mathcal{B}(0,1]$, $x \in E$. Let $k(x,z)$ be the D-component of $k^*(x,z)$ if $k^*(x,z) \in D \times (0,1]$ and let $k(x,z) = \Delta$ if $k^*(x,z) = \Delta$. Then, $\tilde{k}$ and k satisfy (4.33).

There remains to prove (4.34). By a monotone class argument, (4.33) is equivalent to

$$(4.38) \quad \int_0^1 du \int K(x,dy)\ 1_C(y,\tilde{k}(x,y,u)) = \int \nu(dz)\ 1_C(k(x,z),z)$$

holding for all $C \in \mathcal{D} \otimes \mathcal{R}$. Let $C^x = \{\ (y,z) \in D\times\mathbb{R} : k(x,z) \neq y\ \}$. The right-hand-side of (4.38) vanishes when $C = C^x$; hence

$$\int_0^1 du \int K(x,dy)\ I_{\{y\neq k(x,\tilde{k}(x,y,u))\}} = 0,$$

which is equivalent to (4.34).

(4.39) REMARK. In fact, the statement b) of the preceding lemma can be strengthened as follows. If $K(x,\{y\}) = 0$, then we can choose $\tilde{k}(x,y,u) = \hat{k}(x,y)$ free of u and still satisfy (4.33) and (4.34). Hence, $\tilde{k}(x,y,u)$ depends on u only when y is an atom of $K(x,\cdot)$.

(4.40) REMARK. Suppose $K(x,\{y\}) > 0$ and let $B_{xy} = \{ z: k(x,z) = y\}$. Then, $\nu(B_{xy}) = K(x,\{y\}) > 0$. Now, $\tilde{k}(x,y,u) \in B_{xy}$ for Lebesgue-almost every u and

$$\int_0^1 du\, 1_B \circ \tilde{k}(x,y,u) = \nu(B \cap B_{xy})/\nu(B_{xy}).$$

In other words, when $\tilde{k}(x,y,\cdot)$ is regarded as a random variable on the probability space $((0,1],\mathcal{B}(0,1],\text{Leb})$, its distribution is the restriction of ν to B_{xy} normalized. This explains how the construction made below formalizes the recipe (4.9).

We return to the construction problem. Recall the terms defined in §4c: S_{np}, U_{np}, β. Let

$$(4.41)\quad \begin{cases} F_{np} = \{ S_{np}<\infty,\ \beta_{S_{np}} \neq \Delta,\ S_{mq} \neq S_{np} \text{ if } (m,q)\neq(n,p) \}, \\ Z_{np} = \tilde{k}(X_{S_{np}-}, \beta_{S_{np}}, U_{np})\, I_{F_{np}}, \\ \bar{N}(dt,dz) = \sum_{n\in \mathbb{N}^*, p\in \mathbb{N}^*} I_{F_{np}}\, \varepsilon_{(S_{np},Z_{np})}(dt,dz). \end{cases}$$

The random measure $\bar{N}$ is quite close to the Poisson random measure $\tilde{N}$ we are seeking:

(4.42) PROPOSITION. The measure $\bar{N}$ is an additive integer-valued random measure on $(\bar{X},\bar{H}_t)$ whose dual predictable projection is $\bar{G}(dt,dz) = 1_D \circ k(X_t,z)\, dt\, \nu(dz)$.

PROOF. By construction (4.41), $\bar{N}$ satisfies (2.11). That it has the required measurability and additivity properties is easy to check recalling that β is (H_t)-optional and homogeneous and $S_{n1} = S_n$ is a terminal time and (4.11) holds. To prove that $\bar{G}$ is the dual predictable projection of $\bar{N}$ over $(\bar{\Omega},\bar{H},\bar{H}_t,\bar{\mathbb{P}}_x)$, it suffices to show that

$$(4.43) \qquad \bar{\mathbb{E}}_x[\, H\, \bar{N}((a,b]\times B)\,] = \bar{\mathbb{E}}_x[\, H\, \bar{G}((a,b]\times B)\,]$$

for all $0 \le a < b < \infty$ and positive $H \in \bar{H}_a$ and $B \in R$ (then use (2.26) and a monotone class argument). Now, for every $\omega \in \Omega$, $H(\omega,\cdot) \in H_a^\omega$, whereas for every (n,p) such that $S_{np}(\omega) > a$ we have that U_{np} is independent of H_a^ω and has the uniform distribution on $(0,1]$. Hence, by (4.41),

$$(4.44)\ \bar{\mathbb{E}}_x[\, H\, \bar{N}((a,b]\times B)\,]$$

$$= \mathbb{E}_x[\int P^1(dw)H(\cdot,w) \sum_{n,p} I_{F_{np}} 1_{(a,b]}(S_{np}) 1_B \circ \tilde{k}(X_{S_{np}-},\beta_{S_{np}},U_{np})]$$

$$= \mathbb{E}_x[\int P^1(dw)H(\cdot,w) \sum_{n,p} \int_0^1 du\, I_{F_{np}} 1_{(a,b]}(S_{np}) 1_B \circ \tilde{k}(X_{S_{np}-},\beta_{S_{np}},u)]$$

By (4.13) and (4.16), except on a null set, we have $S_{mq} \ne S_{np}$ if $(m,q) \ne (n,p)$, $S_{np} < \infty$, and $\beta_t \ne \Delta$ if and only if there exist (n,p) with $S_{np} = t$. Thus, the last member of (4.44) is equal to

$$(4.45) \quad \int P^1(dw) \int du\, \mathbb{E}_x[\int \Gamma(dt,dy)\, 1_{(a,b]}(t)\, 1_B \circ \tilde{k}(X_{t-},y,u)\, H(\cdot,w)\,].$$

Since $H(\cdot,w) \in \mathcal{H}_a$ for every $w \in \Omega^1$, $(\omega,t) \to H(\omega,w)1_{(a,b]}(t)$ is $(\mathcal{H}_t)$-predictable and $(x,y) \to 1_B \circ \tilde{k}(x,y,u)$ is $\mathcal{E}_0 \otimes \mathcal{D}$-measurable. Thus, using (2.26) and the fact that the dual-predictable projection of Γ on $(\Omega,\mathcal{H},\mathcal{H}_t,\mathbb{P}_x)$ is G given by (3.3), we see that (4.45) is equal to

$$\int P^1(dw) \int_0^1 du\ \mathbb{E}_x [\int_a^b dt \int_D K(X_t,dy)\ 1_B \circ \tilde{k}(X_{t-},y,u)\ H(\cdot,w)].$$

Finally, we can replace X_{t-} by X_t since the two differ for at most countably many t, and using (4.33) we obtain

$$\bar{\mathbb{E}}_x [H\ \bar{N}((a,b]\times B)] = \int P^1(dw)\ \mathbb{E}_x [H(\cdot,w) \int_a^b dt \int_B \nu(dz) 1_D \circ k(X_s,z)]$$

$$= \bar{\mathbb{E}}_x [H\ \bar{G}((a,b]\times B)]$$

as desired.

Finally, recall the auxiliary Poisson random measure N constructed in §4d, and on $\tilde{\Omega}$ set

(4.46) $$\tilde{N}(dt,dz) = \bar{N}(dt,dz) + N(dt,dz) 1_{\{\Delta\}} \circ k(X_{t-},z)\ I_{\{\bar{N}(\{t\}\times\mathbb{R})=0\}}.$$

(4.47) PROPOSITION. The random measure $\tilde{N}$ defined by (4.46) is a Poisson random measure over $(\tilde{X},\tilde{\mathcal{H}}_t)$ with mean measure $dt\ \nu(dz)$. Moreover, (3.9) holds $\tilde{P}_x$-almost surely for all $x \in E$.

PROOF. a) By construction, $\tilde{N}$ is an integer-valued random measure. By (2.11), for each $x \in E$, it is $\tilde{\mathbb{P}}_x$-indistinguishable from a $(\tilde{\mathcal{H}}_t)$-optional random measure not depending on x, and it is additive with respect to $(\tilde{\theta}_t)$ since $\bar{N}$ and N are additive and since X and

$N(\{t\}\times\mathbb{R})$ are homogeneous. Since $\bar{N}$ and N are independent under each $\tilde{\mathbb{P}}_x$ by construction, and since N is Poisson with a diffuse mean measure, almost surely, they have no points occurring simultaneously. Hence,

$$\tilde{N}(dt,dz) = \bar{N}(dt,dz) + N(dt,dz)\, 1_{\{\Delta\}}\circ k(X_{t-},z) \quad \text{a.s.}$$

and the dual predictable projection of $\tilde{N}$ is given by

$$\bar{G}(dt,dz) + dt\, \nu(dz)\, 1_{\{\Delta\}}\circ k(X_{t-},z)$$

by (2.30) and since $1_{\{\Delta\}}\circ k$ is $\mathcal{E}_0 \otimes \mathcal{D}$-measurable. Now using (4.42) we see that the dual predictable projection of $\tilde{N}(dt,dz)$ is $dt\, \nu(dz)$, which by Proposition (2.39) implies that $\tilde{N}$ is a Poisson random measure over $(\tilde{X},\tilde{H}_t)$ as desired.

b) Recall that the S_{np} have their graphs almost surely pairwise disjoint and that $\{ t: \beta_t \neq \Delta \} = \bigcup_{n,p} \{S_{np}\}$ a.s. Let $L = \{ (x,y,u): k(x,\tilde{k}(x,y,u)) \neq y \}$. We have

$$\tilde{\mathbb{E}}_x [\sum_{n,p} 1_L(X_{S_{np}-},\beta_{S_{np}},U_{np})]$$

$$= \sum_{n,p} \int P^1(dw)\, \mathbb{E}_x [1_L(X_{S_{np}-},\beta_{S_{np}},U_{np}(w))]$$

$$= \sum_{n,p} \int_0^1 du\, \mathbb{E}_x [1_L(X_{S_{np}-},\beta_{S_{np}},u)]$$

$$= \int_0^1 du\, \mathbb{E}_x [\int \Gamma(dt,dy)\, 1_L(X_{t-},y,u)]$$

$$= \int_0^1 du\, \mathbb{E}_x [\int G(dt,dy)\, 1_L(X_{t-},y,u)]$$

$$= \mathbb{E}_x [\int_0^\infty dt \int_0^1 du \int_D K(X_t,dy)\, 1_L(X_t,y,u)],$$

and this last expression equals 0 by (4.34). Hence, if $B \in \mathcal{R}_+ \otimes \mathcal{D}$,

$$\int \tilde{N}(dt,dz)\, 1_N(t,k(X_{t-},z))$$

$$= \sum_{n,p} 1_B(S_{np},k(X_{S_{np}-},\tilde{k}(X_{S_{np}-},\beta_{S_{np}},U_{np})))\, 1_{L^c}(X_{S_{np}-},\beta_{S_{np}},U_{np})$$

$$= \sum_{n,p} 1_B(S_{np},\beta_{S_{np}}) = \Gamma(B)$$

almost surely. That is, (3.9) holds almost surely.

(4.48) REMARK. Assume $K(x,\cdot)$ is diffuse and infinite for all $x \in E$. Proposition (4.33b) implies that $\tilde{k}(x,y,u) = \hat{k}(x,y)$ for all y for some $\hat{k}$. Hence, in (4.41), the variables U_{np} do not intervene and we need not add the space Ω^1. Also, in the proof of Proposition (4.33), in step 4, it is easy to see that we may choose the G_n^x so that $\bigcup_n G_n^x = \mathbb{R}$ by using the fact that $K(x,D) = \infty$ and ν is σ-finite. Then, $k(x,z)$ never takes the value Δ, and in (4.16), N does not intervene; hence we do not need to add the space Ω^2. Thus, we need no enlargements and we have $\tilde{X} = X$ and $\tilde{\mathcal{H}}_t = \mathcal{H}_t$.

References

[1] A. BENVENISTE et J. JACOD. Systèmes de Lévy des processus de Markov. *Invent. Math.* *21* (1973), 183-198.

[2] R.M. BLUMENTHAL and R.K. GETOOR. *Markov Processes and Potential Theory*. Academic Press, New York, 1968.

[3] E. ÇINLAR. Markov additive processes, II. *Z. Wahrscheinlichkeitstheorie verw. Gebiete*, *24* (1972), 94-121.

[4] E. ÇINLAR and J. JACOD. After a time change every Hunt process satisfies a stochastic integral equation driven by a Wiener process and a Poisson random measure. To appear.

[5] E. ÇINLAR, J. JACOD, P. PROTTER, and M.J. SHARPE. Semimartingales and Markov processes. *Z. Wahrscheinlichkeitstheorie verw. Gebiete 54* (1980), 161-219.

[6] E.B. DYNKIN. *Markov Processes.* Academic Press, New York, 1965.

[7] N. EL KAROUI and J.-P. LEPELTIER. Représentation des processus ponctuels multivariés à l'aide d'un processus de Poisson. *Z. Wahrscheinlichkeitstheorie verw. Gebiete, 39* (1977), 111-133.

[8] W. FELLER. The general diffusion operator and positivity preserving semigroups in one dimension. *Ann. Math. 60* (1954), 417-436.

[9] W. FELLER. On second order differential operators. *Ann. Math. 61* (1955), 90-105.

[10] W. FELLER. Generalized second order differential operators and their lateral conditions. *Illinois J. Math 1* (1957), 495-504.

[11] R.K. GETOOR. *Markov Processes: Ray Processes and Right Processes.* Lecture Notes in Math. *440*, Springer-Verlag, Berlin, 1975.

[12] B. GRIGELIONIS. On the representation of integer-valued random measures by means of stochastic integrals with respect to the Poisson measure. *Lit. Math. J. 11* (1971), 93-108.

[13] K. ITO. On stochastic processes (I) (Infinitely divisible laws of probability). *Japan J. Math. 18* (1942), 261-301.

[14] K. ITO. On Stochastic Differential Equations. *Mem. Amer. Math. Soc. 4* (1951).

[15] J. JACOD. Fonctionnelles additives et systèmes de Lévy des produits semi-directs de processus de Markov. *Bull. Soc. Math. France, 35* (1973), 119-144.

[16] J. JACOD. *Calcul Stochastique et Problèmes de Martingales.* Lecture Notes in Math. *714*, Springer-Verlag, Berlin, 1979.

[17] Yu.M. KABANOV, R.S. LIPTSER, A.N. SHIRYAYEV. On the representation of integer-valued random measures and of local martingales with respect to random measures with deterministic compensators. *Math. Sb. 111* (1980), 293-307.

[18] F.B. KNIGHT. An infinitesimal decomposition for a class of Markov processes. *Ann. Math. Statist. 41* (1970), 1510-1529.

[19] H. KUNITA and S. WATANABE. On square integrable martingales. *Nagoya J. Math. 30* (1967), 209-245.

[20] P.A. MEYER. Integrales stochastiques III. *Séminaire de Probabilités I (Univ. Strasbourg)*, pp. 118-141. Lecture Notes Math. *39*, Springer-Verlag, Berlin, 1967.

[21] A.V. SKOROKHOD. On homogeneous continuous Markov processes that are martingales. *Theo. Prob. Appl. 8* (1963), 355-365.

[22] A.V. SKOROKHOD. On the local structure of continuous Markov processes. *Theo. Prob. Appl. 11* (1966), 366-372.

[23] D.W. STROOCK. Diffusion processes associated with Lévy generators. *Z. Wahrscheinlichkeitstheorie verw. Gebiete, 32* (1975), 209-244.

[24] D.W. STROOCK and S.R.S. VARADHAN. *Multidimensional Diffusion Processes.* Springer-Verlag, Berlin, 1979.

[25] H. TANAKA. Probabilistic treatment of the Boltzmann equation of Maxwellian molecules. *Z. Wahrscheinlichkeitstheorie verw. Gebiete, 46* (1978), 67-105.

[26] J.B. WALSH. The perfection of multiplicative functionals. *Séminaire de Probabilités VI (Univ. Strasbourg)*, pp. 233-242. Lecture Notes in Math. *258*, Springer-Verlag, Berlin, 1972.

[27] M. YOR. Un exemple de processus qui n'est pas une semimartingale. *Temps Locaux*, pp. 219-222. *Astérisque, No. 52-53* (1978).

IE/MS Department
Northwestern University
Evanston, IL 60201 U.S.A.

IRISA, Lab. associé au CNRS
Université de Rennes
F-35031 Rennes Cedex, France